D1116749

Composite Materials for Aircraft Structures

Edited by
B.C. Hoskin and A.A. Baker
Aeronautical Research Laboratories
Melbourne, Victoria, Australia

AIAA EDUCATION SERIES
J. S. Przemieniecki
Series Editor-in-Chief
Air Force Institute of Technology
Wright-Patterson Air Force Base, Ohio

Published by
American Institute of Aeronautics and Astronautics, Inc.
1633 Broadway, New York, N.Y. 10019

Texts Published in the AIAA Education Series

Re-Entry Vehicle Dynamics
Frank J. Regan, 1984
Aerothermodynamics of Gas Turbine and Rocket Propulsion
Gordon C. Oates, 1984
Aerothermodynamics of Aircraft Engine Components
Gordon C. Oates, Editor, 1985
Aircraft Combat Survivability Analysis and Design
Robert E. Ball, 1985
Intake Aerodynamics
J. Seddon and E.L. Goldsmith, 1985

American Institute of Aeronautics and Astronautics, Inc., Washington, DC

Library of Congress Cataloging in Publication Data

Hoskin, B.C. and Baker, A.A.
Composite materials for aircraft structures.

(AIAA education series)
Includes index.
1. Airplanes—Materials. 2. Composite materials.
I. Hoskin, B.C. II. Baker, A.A. (Alan A.) III. Series.
TL699.C57C66 1986 629.134 86-1244
ISBN 0-930403-11-8

Fifth Printing

Foreword

Composite Materials for Aircraft Structures, edited by B. C. Hoskin and A. A. Baker, is the latest addition to the AIAA Education Series inaugurated in 1984. The series represents AIAA's response to the need for textbooks and monographs in highly specialized disciplines of aeronautics and astronautics. *Composite Materials for Aircraft Structures*, just such a case in point, should prove particularly timely because the field has surged in composite applications.

Composite Materials for Aircraft Structures provides a broad introduction to virtually all aspects of the technology of composite materials for aircraft structural applications: the basic theory of fiber reinforcements; material characteristics of the commonly used fibers, resins, and composite systems; component form and manufacture; structural mechanics of composite laminates; composite joints; environmental effects; durability and damage tolerance; nondestructive inspection (NDI) and repair procedures; aircraft applications; and airworthiness considerations.

This text, expanded and updated, has been prepared from notes used in a series of lectures given at the Aeronautical Research Laboratories (ARL), Melbourne, Victoria, Australia. All lecturers were officers in either the Structures or Aircraft Materials Divisions of ARL. The table of contents gives the names of the lecturers, together with their topics.

The lectures originated with a request to ARL from the Australian Department of Aviation's Airworthiness Branch. The Director of the Aeronautical Research Laboratories, Department of Defense, Australia, has authorized publication of the expanded and updated text by AIAA.

J. S. PRZEMIENIECKI
Editor-in-Chief
AIAA Education Series

Table of Contents

ix **Preface**

1 **Chapter 1. Introduction,** *B.C. Hoskin*
- 1.1 General
- 1.2 Overview of Use of Composite Materials in Aircraft Structures
- 1.3 Composite Materials for Other Aeronautical Applications
- 1.4 Bibliographic Comments

9 **Chapter 2. Basic Principles of Fiber Composite Materials,** *A.A. Baker and A.W. Rachinger*
- 2.1 General
- 2.2 The Elastic Constants
- 2.3 Mechanics of Materials Approach to Strength
- 2.4 Fracture Toughness

37 **Chapter 3. Fiber Systems,** *J.G. Williams*
- 3.1 Introduction
- 3.2 Manufacture of Commercial Fibers
- 3.3 Forms of Fiber
- 3.4 Sizes, Finishes, and Fiber Surface Treatments

47 **Chapter 4. Resin Systems,** *J.G. Williams*
- 4.1 Introduction
- 4.2 Epoxy Resins
- 4.3 Polyester Resins
- 4.4 Vinyl-Ester Resins
- 4.5 Phenolic Resins
- 4.6 Other Resins
- 4.7 Carbon-Carbon Composites
- 4.8 Fiber-Reinforced Thermoplastics
- 4.9 Quality Control of Resins, Resin Systems, and Prepregs

59 **Chapter 5. Composite Systems,** *M.J. Davis and A.W. Rachinger*
- 5.1 Introduction
- 5.2 Available Forms of Material
- 5.3 Properties of Advanced Composite Systems

5.4 Characterization of Composites
5.5 The Mechanical Testing of Composites
5.6 Hybrid Composites

73 **Chapter 6. Component Form and Manufacture,** *A.A. Baker*
6.1 Introduction
6.2 An Outline of General Laminating Procedures
6.3 Laminating Procedures for Advanced Fiber Composites
6.4 Filament Winding
6.5 Pultrusion
6.6 Machining

93 **Chapter 7. Structural Mechanics of Fiber Composites,** *B.C. Hoskin and B.I. Green*
7.1 Introduction
7.2 Stress-Strain Law for Single Ply in Material Axes: Unidirectional Laminates
7.3 Stress-Strain Law for Single Ply in Structural Axes: Off-Axis Laminates
7.4 Plane Stress Problems for Symmetric Laminates
7.5 General Laminates Subjected to Plane Stress and Bending Loads
7.6 Bending of Symmetric Laminates
7.7 Failure Criteria for Laminates

115 **Chapter 8. Joining Advanced Fiber Composites,** *A.A. Baker*
8.1 Introduction
8.2 Adhesive Bonding
8.3 Mechanical Joints

141 **Chapter 9. Environmental Effects and Durability,** *B.C. Hoskin*
9.1 Introduction
9.2 Moisture Absorption
9.3 Effect of Moisture and Temperature on Structural Performance
9.4 Other Environmental Effects
9.5 Fatigue of Composites in Normal Environment
9.6 Further Comments on Composite Fatigue
9.7 Final Remarks

153 **Chapter 10. Damage Tolerance of Fiber Composite Laminates,** *M.J. Davis and R. Jones*
10.1 Introduction
10.2 Nature of Defects
10.3 Damage-Tolerant Design
10.4 Damage Tolerance Improvement

175 **Chapter 11. NDI of Fiber-Reinforced Composite Materials,** *I.G. Scott and C.M. Scala*

11.1 Introduction
11.2 The Search for Defects
11.3 Attempts to Assess Structural Integrity
11.4 Summary

193 **Chapter 12. Repair of Graphite/Epoxy Composites,** *A.A. Baker*

12.1 Introduction
12.2 Damage Assessment
12.3 General Repair Approaches
12.4 Material Aspects
12.5 Design of Bonded Repairs
12.6 Certification of Repairs

217 **Chapter 13. Aircraft Applications,** *B.C. Hoskin and A.A. Baker*

13.1 Introduction
13.2 Typical Composite Constructions
13.3 Design Principles for Composite Structure
13.4 Erosion Protection
13.5 Lightning Protection
13.6 Helicopter Applications

225 **Chapter 14. Airworthiness Considerations,** *B.C. Hoskin*

14.1 Introduction
14.2 Demonstration of Static Strength
14.3 Demonstration of Fatigue Strength
14.4 Damage Tolerance
14.5 A Proposed Schedule of Environmental Tests for Composite Certification
14.6 Flammability
14.7 Force Management

235 **Subject Index**

Preface

The use of composite materials for aircraft structural applications has been expanding rapidly in recent years. However, in virtually all aspects, the use of these materials involves a very different technology from that for metals. The materials themselves are intrinsically different; so too are manufacturing processes, structural design procedures, and the in-service performance of the materials, particularly regarding the cause and nature of damage that may be sustained. All of these important matters are addressed in this textbook on the technology of composite materials for aircraft structural applications.

The opening chapters of *Composite Materials for Aircraft Structures* set the scene of modern engineering with composites and describe basic principles of fiber composites, a general account being given of how the properties of the heterogeneous composite derive from the properties of its constituents, fibers plus matrix—the "micromechanics" of composites.

The next three chapters deal with the physical and mechanical properties of composite materials; the properties of the commonly used fibers, in particular, graphite, boron, aramid, and glass; the resins used as matrix materials, especially the widely used epoxy resins; and properties of the main composites—graphite/epoxy, boron/epoxy, aramid/epoxy, and glass/epoxy.

Chapter 6 outlines procedures involved in manufacturing a composite component, particular attention being paid to the use of tape or broadgoods, but some reference also is made to filament winding.

The next four chapters deal with the structural applications of composites. Chapter 7 presents the basic theory needed for composite structural analysis. This involves concepts from anisotropic elasticity. Mechanical properties of composites are highly dependent directionally; and, indeed, the key to the successful application of composites lies in the proper utilization of this directional dependence. Basics also draw on laminate theory because composite structures are usually made as multi-ply laminates. Chapter 8 takes up the especially important topic of joints, both mechanically fastened and adhesively bonded. Chapter 9 presents factors affecting in-service performance and durability, especially environmental effects (e.g., moisture absorption and high temperature) and fatigue. Chapter 10 describes damage tolerance, especially for damage caused by low-energy impacts.

The next two chapters go into nondestructive inspection (NDI) methods for use in the factory and field and procedures used for repairing composite structures, again both for the factory and the field.

Finally, Chapters 13 and 14 cover types of composite construction being used in current aircraft and matters bearing on airworthiness certification of aircraft containing major composite components.

This textbook will, we believe, contribute to the overall understanding of the use of composite materials in aircraft structures, and help lead to much more extensive applications of composites in future designs.

B. C. HOSKIN
A. A. BAKER
Aeronautical Research Laboratories
Melbourne, Australia

1. INTRODUCTION

1.1 GENERAL

A major development in aeronautics has come to the fore over the last decade or so—the use of composite materials in place of metals in aircraft structures. In general, a composite material can be defined simply as a material that consists of two (or more) identifiably distinct constituent materials. The composite materials used for aircraft structures belong to the class known as "fiber composites" (or, sometimes, "fiber reinforced plastics") comprising continuous fibers embedded in a resin (or "plastic") matrix. It is the fibers that provide such a composite with its key structural properties, the matrix serving mainly to bond the fibers into a structural entity. The prime reason for using composite materials is that substantial weight savings can be achieved because of their superior strength-to-weight and stiffness-to-weight ratios, as compared with the conventional materials of aircraft construction such as the aluminum alloys. Weight savings of the order of 25% are generally considered to be achievable using current composites in place of metals.

There is a distinct lack of uniformity in the names given to composite materials. In the United States, the usual practice is to write the name in the format "fiber/matrix." Since, for aircraft structural applications, the main fibers presently used are graphite, boron, aramid, and glass and the main matrix material is epoxy resin, the main composites in the U.S. terminology are graphite/epoxy, boron/epoxy, aramid/epoxy, and glass/epoxy. (Naturally, if another type of matrix material is used, that is reflected in the name; for example, there has been some interest in graphite/polyimide composites.) This notation is reasonably explicit and a further advantage is that it can be readily adapted to describe specific composite systems. For example, a common U.S. graphite/epoxy system uses Thornel T300 fibers and Narmco 5208 resin; this is generally abbreviated to T300/5208. Another common U.S. graphite/epoxy system uses Hercules AS fibers and Hercules 3501-6 resin; this is likewise abbreviated to AS/3501-6. The American terminology will be adopted here, but mention will be made of the main alternatives. In the United Kingdom, where graphite fibers were first developed, they have always been called carbon fibers and the associated composite is simply called carbon fiber composite. Aramid fibers are organic fibers first developed by Du Pont and the Du Pont proprietary name of Kevlar is commonly used; originally, the same material was known as PRD-49. Finally, at one time, the terms carbon fiber-reinforced plastic, boron fiber-reinforced plastic, and glass fiber-reinforced plastic were in wide use; this terminology seems to be lapsing, partly because it suggests that the fibers act in a supporting role to the plastic matrix, whereas the real situation is quite the reverse. The terminology has been summarized in Table 1.1.

1.2 OVERVIEW OF USE OF COMPOSITE MATERIALS IN AIRCRAFT STRUCTURES

Of the materials being considered here, glass fiber composites were the first to be used for aircraft structures. As far back as 1944, a Vultee BT-15 trainer aircraft was made and flown with the aft fuselage skin made of glass fiber composite sandwich panels (composite facings on a balsa wood core). However, while over the succeeding years glass fiber composites, especially in the form of glass/epoxy, have become quite widely used in aircraft structures, this use has been mainly confined to items such as control surfaces, fairings, interior fittings, and canopies; this is the way glass fiber composites are utilized in, for example, the Boeing 747. The reason why these composites have not been generally employed for major structural components is that, although their strength-to-weight ratio compares very favorably with that for metals, their stiffness-to-weight ratio does not and stiffness is often as important a design requirement as strength, especially for high-speed aircraft. One application of glass/epoxy that is being widely pursued, however, is in helicopter rotor blades.

The potential for the widespread use of composites in aircraft structures came about with the more or less simultaneous invention, around 1960, of graphite fibers in the UK and boron fibers in the U.S. The so-called "advanced composite materials" based on either of these fiber types, generally with an epoxy matrix, are markedly superior to conventional aircraft materials in both strength and stiffness properties. Initially, the development of boron/epoxy proceeded the more rapidly and virtually entirely in the U.S. By 1970, the U.S. had made and flown major boron/epoxy demonstrator items, including an F-111 horizontal tail and 50 F-4 rudders. On the basis of the experience thus gained, boron/epoxy was incorporated in the U.S. high-performance military aircraft then being designed. Thus, the skin of the horizontal tail of the F-14 and of both the horizontal and vertical tails of the F-15 are made of boron/epoxy. The development of graphite/epoxy went on much more slowly in the UK;

Table 1.1 Main Composite Materials for Aircraft Structural Applications

Nomenclature Used in this Text	Equivalent Nomenclature
Graphite/epoxy (Gr/Ep)	Carbon fiber composite (CFC) Carbon fiber-reinforced plastic (CFRP)
Boron/epoxy (B/Ep)	Boron fiber-reinforced plastic (BFRP)
Aramid/epoxy (Ar/Ep)	Kevlar/epoxy PRD-49/epoxy
Glass/epoxy	Glass fiber-reinforced plastic (GFRP or GRP)

generally, only small demonstrator items such as the rudder trim tabs for a Strikemaster and a spoiler for the Jaguar were being made at that time.

However, by the mid-1970s, the U.S. had decided to switch from boron/epoxy to graphite/epoxy. The reason was primarily the cost of the material; by 1979 the cost of graphite/epoxy in the U.S. in the form in which it is procured by the aircraft manufacturer (namely, "prepreg") was \$40/lb, while that of boron/epoxy was \$180/lb. Having made the switch, the U.S. has been quick to incorporate graphite/epoxy in its high-performance military aircraft. In the F-16, graphite/epoxy is used for the skin of the horizontal and vertical tails and for various control surfaces; it comprises about 3% of the structural weight. A more extensive use of graphite/epoxy is made in the F/A-18 (Fig. 1.1); there the wing skins, horizontal and vertical tail skins, fuselage dorsal cover and avionics bay door, speed brake, and many of the control surfaces are graphite/epoxy, comprising 9% of the structural weight (and 35% of the surface area). In the AV-8B (Fig. 1.2), almost the complete wing, i.e., skin plus substructure, is made of graphite/epoxy; it is also used in the horizontal tail, forward fuselage and various control surfaces and comprises about 26% of the structural weight. The horizontal tail (skin and substructure) of a prototype B-1 bomber was graphite/epoxy. The wing skins of the X-29 forward swept wing demonstration aircraft are graphite/epoxy; here, the anisotropic nature of the fiber composites is utilized to minimize the torsional divergence problem.

With regard to European military aircraft, a graphite/epoxy taileron (i.e., all-moving horizontal tail) has been developed for the Tornado by the UK

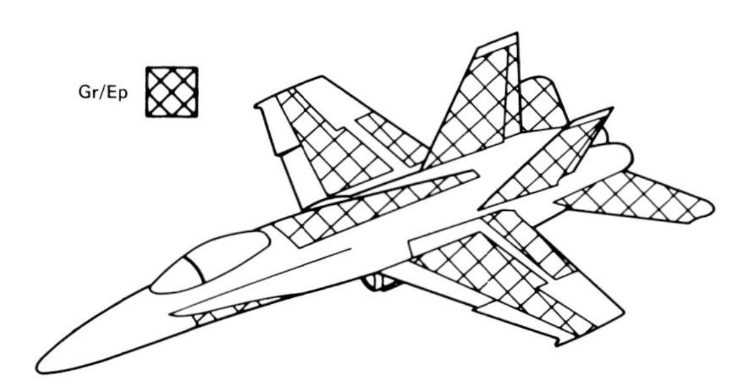

Fig. 1.1 Graphite/epoxy applications on the F/A-18.

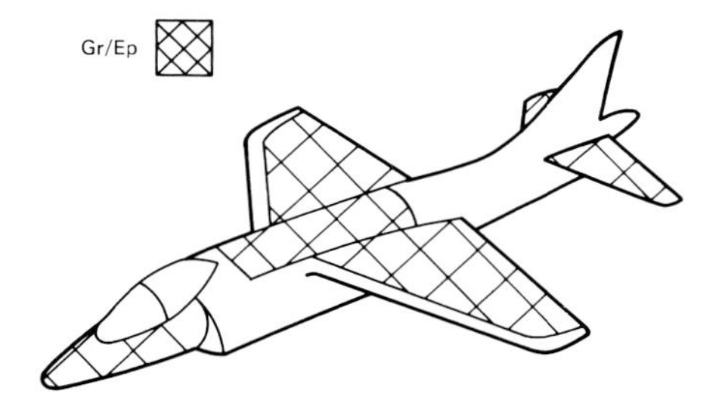

Fig. 1.2 Use of graphite/epoxy in the AV-8B (Harrier).

and West Germany and the French Mirage 2000 uses both boron and graphite composites for its tail unit and control surfaces.

Applications of advanced composites in civil aircraft have lagged behind those in military aircraft; however, interest is now quickening. Various graphite/epoxy demonstrator items have been made in the U.S. Early such items included the Boeing 737 spoilers, of which 111 were made and fitted to the aircraft of seven airlines operating throughout the world; by mid-1981, one spoiler had achieved approximately 22,000 flight hours and no significant problems had been encountered. The substantial weight savings that can be achieved using composites can lead to substantial fuel savings and, because of this, a major development of composite aircraft structures has been undertaken within the framework of the NASA Aircraft Energy Efficiency (ACEE) program. The aim of this development was to provide the technology and confidence so that commercial transport manufacturers can commit themselves to the use of composites in their future aircraft; the time scale is for composites to be used for secondary structure from 1980 onward and for primary structure from 1985 onward. Demonstrator items included in the ACEE composites program are shown in Table 1.2 along with the savings in weight as compared with existing metal components.

Consistently with the above time scales, the Boeing 757 and 767 aircraft use graphite/epoxy for many of the control surfaces (Fig 1.3). Special mention should be made of the Lear Fan 2100 (Fig. 1.4), which is a small passenger aircraft that is sometimes referred to as the "all-composite" aircraft because almost all of the airframe is made of composites, mainly graphite/epoxy. The Rutan Voyager two-seat aircraft, designed for a non-stop around-the-world flight, has a virtually all-composite structure.

Composites have also been used in space vehicle structures; e.g., the 15 m long cargo bay doors in the Space Shuttle are graphite/epoxy. There have been many applications of graphite/epoxy in satellite structures where another advantage of the material (besides the weight savings) is its low coefficient of thermal expansion, which permits the maintenance of dimensional stability under large temperature variations.

So far no mention has been made of aramid/epoxy, another advanced composite. This is quite widely used in aircraft structures, but largely in roles previously filled by glass/epoxy. However, there is interest in a hybrid

Table 1.2 Graphite/Epoxy Demonstrators in ACEE Program (see p. 8 of Ref. 9)

Type of Structure	Component	Weight Saving, %
Secondary	McDonnell-Douglas DC-10 rudder	26.8
	Boeing 727 elevator	25.6
	Lockheed L-1011 aileron	26.3
Primary	McDonnell-Douglas DC-10 vertical tail	20.2
	Boeing 737 horizontal tail	27.1
	Lockheed L-1011 vertical tail	27.9

aramid-graphite composite for more general use. In the Boeing 767, for example, this hybrid is used for the wing-to-fuselage fairing, undercarriage doors, engine cowlings, and fixed trailing-edge panels. One shortcoming of aramid composites militating against their wider use in primary structures has been their rather low compressive strength. (However, this problem is ameliorated in the hybrid aramid-graphite composite.)

In summary, fiber composite materials are already being used to a significant extent in aircraft structures and that use seems certain to be extended. A consolidated list of some major applications is given in Table 1.3. Currently, graphite/epoxy and, to a lesser extent, aramid/epoxy are seen as the most important composites for aircraft structural applications.

1.3 COMPOSITE MATERIALS FOR OTHER AERONAUTICAL APPLICATIONS

While this text is concerned with composite materials for aircraft structures, other types of composites are being investigated for a variety of possible aeronautical applications. One particularly active area is in the development of metal matrix composites for use in, for example, jet engine

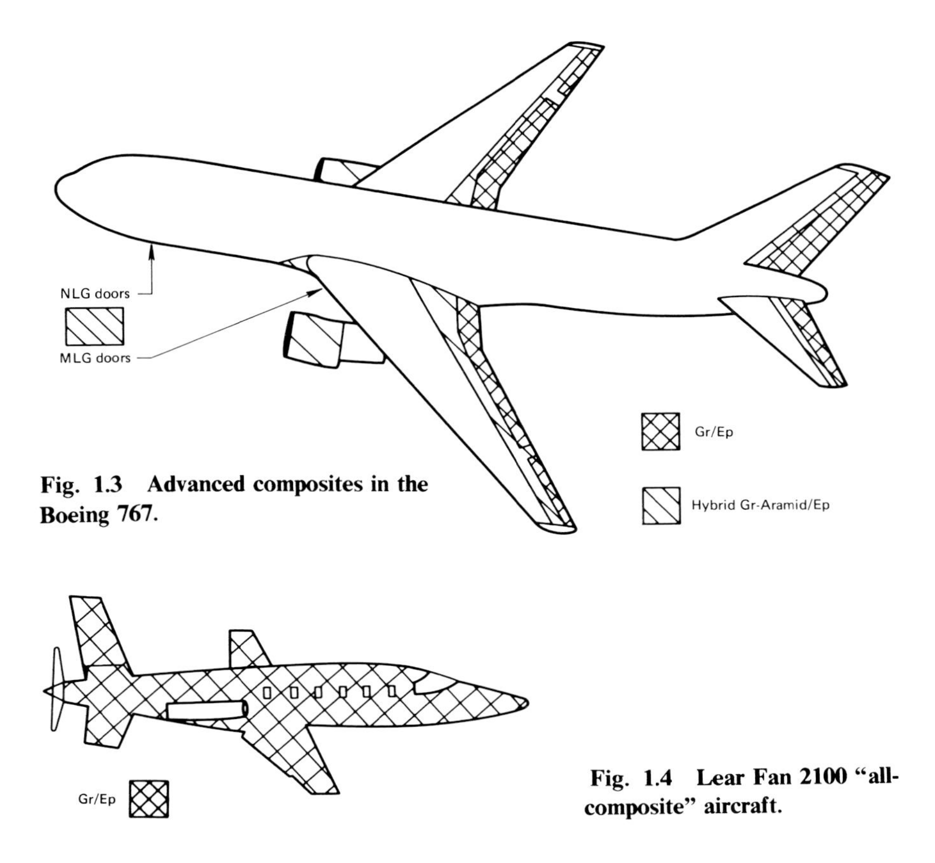

Fig. 1.3 Advanced composites in the Boeing 767.

Fig. 1.4 Lear Fan 2100 "all-composite" aircraft.

Table 1.3 Some Aircraft Applications of Advanced Composites

Aircraft	Composite	Application
F-14	B/Ep	Horizontal tail skin
F-15	B/Ep	Horizontal tail skin
	B/Ep	Vertical tail skin
	Gr/Ep	Speed brake
F-16	Gr/Ep	Horizontal tail skin
	Gr/Ep	Vertical tail skin
	Gr/Ep	Control surfaces
F/A-18	Gr/Ep	Wing skin
	Gr/Ep	Horizontal tail skin
	Gr/Ep	Vertical tail skin
	Gr/Ep	Control surfaces, speed brake
	Gr/Ep	Dorsal cover, avionics bay door
AV-8B	Gr/Ep	Wing skin and substructure
	Gr/Ep	Horizontal tail skin
	Gr/Ep	Forward fuselage
	Gr/Ep	Control surfaces
X-29	Gr/Ep	Wing skin
Boeing 757 & 767	Gr/Ep	Control surfaces
	Ar-Gr/Ep	Fairings, undercarriage doors, cowlings
Lear Fan 2100	Gr/Ep	"Almost all" of structure

components. For instance, boron/aluminum composites (i.e., boron fibers in an aluminum alloy matrix) are being studied for use in fan blades). For hot end components such as turbine blades, there is interest in tungsten/superalloy composites.

Composites not based on continuous fiber reinforcement, but rather comprising particles of one material embedded in a matrix of another material—the so-called "particulate composites"—are also receiving attention. (The most homely of all composites, namely, concrete, is of this form, consisting of particles of stone in a cement matrix. Reinforced concrete is, of course, also a composite, but it is akin to a fiber composite with the steel reinforcing rods playing the part of the fibers.) There is also interest in whisker composites, where the reinforcement is supplied by very small crystals (or "whiskers"); however, these are still largely in a developmental stage.

1.4 BIBLIOGRAPHIC COMMENTS

There is an extensive, and rapidly growing, literature on composite materials. Only some of the more general references will be mentioned here.

Reasonably broad, yet still reasonably succinct, accounts of the field have been given by Jones[1] and Aggarwal and Broutman.[2] The most ambitious work to date is the eight-volume treatise edited by Broutman and Krock.[3]

The U.S. Military Handbook[4] contains an extensive discussion on composites for aerospace applications, although the materials data included in the 1971 edition are limited to glass and boron composites. The several publications by the NATO Advisory Group for Aerospace Research and Development (AGARD) provide a useful background on the way in which composites have been developed for aeronautical applications.[5-8] Several specific applications of composites in aircraft structures are described in detail in Refs. 9 and 10.

Some idea of the many aspects of composites on which research is being undertaken can be obtained by consulting the conference proceedings regularly published by the American Society for Testing and Materials (ASTM).[11-15] The state-of-the-art review by Gerharz and Schutz[16] contains an extensive bibliography. Various other items are listed as Refs. 17–26.

References

[1]Jones, R. M., *Mechanics of Composite Materials*, McGraw-Hill Kogakusha Ltd., Tokyo, 1975.

[2]Aggarwal, B. D. and Broutman, L. J., *Analysis and Performance of Fiber Composites*, John Wiley & Sons, New York, 1980.

[3]Broutman, L. J. and Krock, R. H. (eds.), *Composite Materials*, Vols. 1–8, Interscience Publishers, New York, 1975.

[4]*Plastics for Aerospace Vehicles*, Pt. 1, "Reinforced Plastics," MIL-HDBK-17A, U.S. Department of Defense, Washington, DC, 1971.

[5]"The Potentials of Composite Structures in the Design of Aircraft," AGARD Advisory Rept. 10, 1967.

[6]*Composite Materials*, AGARD Conference Proceedings 63, 1970.

[7]Rosen, B. W. (ed.), *Composite Materials*, AGARD Lecture Series, No. 55, 1972.

[8]"Certification Procedures for Composite Structures," AGARD Rept. 660, 1977.

[9]Lenoe, E. M., Oplinger, D. W., and Burke, J. J. (eds.), *Fibrous Composites in Structural Design*, Plenum Press, New York, 1980.

[10]"The 1980s—Pay-Off Decade for Advanced Materials," *Science of Advanced Materials and Process Engineering Series*, Vol. 25, Society for the Advancement of Material and Process Engineering (SAMPE), 1980.

[11]*Composite Materials: Testing and Design*, ASTM STP 460, 1969.

[12]*Composite Materials: Testing and Design (Second Conference)*, ASTM STP 497, 1972.

[13]*Composite Materials: Testing and Design (Third Conference)*, ASTM STP 546, 1974.

[14]*Composite Materials: Testing and Design (Fourth Conference)*, ASTM STP 617, 1977.

[15]Tsai, S. W. (ed.), *Composite Materials: Testing and Design (Fifth Conference)*, ASTM STP 674, 1979.

[16]Gerharz, J. J. and Schutz, D., "Literature Research on the Mechanical Properties of Fibre Composite Materials—Analysis of the State of the Art," Vol. 1, Royal Aircraft Establishment, Farnborough, Library Translation 2045, Aug. 1980.

[17]Tsai, S. W. and Hahn, H. T., *Introduction to Composite Materials*, Technomic Publishing Co., Westport, CT, 1980.

[18]Schwartz, R. T. and Schwartz, H. S. (eds.), *Fundamental Aspects of Fiber Reinforced Plastic Components*, Interscience Publishers, New York, 1968.

[19]Tsai, S. W., Halpin, J. C., and Pagano, N. J. (eds.), *Composite Materials Workshop*, Technomic Publishing Co., Westport, CT, 1968.

[20]Wendt, F. W., Liebowitz, H., and Perrone, N. (eds.), *Mechanics of Composite Materials, Proceedings of Fifth Symposium on Naval Structural Mechanics*, Pergamon Press, Oxford, England, 1970.

[21]Dietz, A. G. H. (ed.), *Composite Engineering Laminates*, MIT Press, Cambridge, MA, 1969.

[22]Gill, R. M., *Carbon Fibres in Composite Materials*, Iliffe, London, 1972.

[23]Salkind, M. J. and Holister, G. S. (eds.), *Applications of Composite Materials*, ASTM STP 524, 1973.

[24]Lubin, G. (ed.), *Handbook of Fiberglass and Advanced Plastics Composites*, Van Nostrand Reinhold, New York, 1969.

[25]Vinson, J. R. and Chou T-W, *Composite Materials and Their Use in Structures*, Applied Science Publishers, London, 1975.

[26]Langley, M. (ed.), *Carbon Fibres in Engineering*, McGraw-Hill Book Co., London, 1973.

2. BASIC PRINCIPLES OF FIBER COMPOSITE MATERIALS

2.1 GENERAL

Introduction to Fiber Composite Systems

A fiber composite material usually consists of one or more filamentary phases embedded in a continuous matrix phase. The aspect ratio (i.e., ratio of length to diameter) of the filaments may vary from about 10 to infinity (for continuous fibers). Their scale, in relation to the bulk material, may range from microscopic (e.g., 8 μm diameter graphite fibers in an epoxy matrix) to gross macroscopic (e.g., 25 mm diameter steel tendons in concrete). This book is concerned mainly with continuous fibers having diameters in the microscopic range.

Composite constituents (fibers and matrices) can be conveniently classified according to their moduli and ductilities. Within the composite, the fibers may, in general, be in the form of continuous fibers, discontinuous fibers, or whiskers (very fine single crystals with lengths of the order of 100–1000 μm and diameters of the order of 1–10 μm) and may be aligned to varying degrees or randomly oriented. This classification is illustrated in Fig. 2.1 for a number of common fibers and matrices. Several examples of composites formed from these materials are shown. The systems of particular interest here are in the brittle fiber/brittle matrix category with the fibers being very much stiffer and stronger than the matrix. These include aramid/epoxy, boron/epoxy, glass/epoxy, and graphite/epoxy.

Micromechanical vs Macromechanical View of Composites

Fiber composites can be studied from two points of view: micromechanics and macromechanics. Micromechanical analyses are aimed at providing an understanding of the behavior of composites (generally unidirectional composites) in terms of the properties and interactions of the fibers and matrix. Approximate models are used to simulate the microstructure of the composite and hence predict its "average" properties (such as strength and stiffness) in terms of the properties and behavior of the constituents.

Advanced composite structures are usually in the form of unidirectional plies* laminated together at various orientations or, alternatively, filament-

*A unidirectional ply is one in which all of the fibers are aligned in the one direction.

wound configurations. In aircraft applications, the laminated multidirectional lay-up is the most common. Macromechanics is used to design, or predict, the behavior of such structures on the basis of the "average" properties of the unidirectional material: namely, the longitudinal modulus E_1, transverse modulus E_2, major Poisson's ratio ν_{12}, and the in-plane shear modulus G_{12}, as well as the appropriate strength values.

Micromechanics is aimed at predicting and understanding these "average" properties in terms of the detailed microscopic behavior of the material, rather than generating accurate design data; macromechanics draws mainly on the results obtained from physical and mechanical testing of unidirectional composites.

Micromechanics

As already mentioned, micromechanics utilizes microscopic models of composites, in which the fibers and the matrix are separately modeled. In most simple models, the fibers are assumed to be homogeneous, linearly elastic, isotropic, regularly spaced, perfectly aligned, and of uniform length. The matrix is assumed to be homogeneous, linearly elastic, and isotropic. The fiber/matrix interface is assumed to be perfect, with no voids or disbonds.

More complex models, representing more realistic situations, may include voids, disbonds, flawed fibers (including statistical variations in flaw severity), wavy fibers, nonuniform fiber dispersions, fiber length variations, and residual stresses.

Micromechanics can, itself, be approached from two points of view:

(1) "The mechanics of materials" approach, which attempts to predict the behavior of simplified models of the composite material.

(2) "The theory of elasticity" approach, which is often aimed at producing upper and lower bound solutions, exact solutions for very specific systems, or numerical solutions.

A common aim of both of these approaches is to determine the elastic constants and strengths of composites in terms of their constituent properties. As previously stated, the main elastic constants for unidirectional fiber composites are:

E_1 = longitudinal modulus (i.e., modulus in fiber direction)

E_2 = transverse modulus

ν_{12} = major Poisson's ratio (i.e., ratio of contraction in the transverse direction consequent on an extension in the fiber direction)

G_{12} = in-plane shear modulus

The main strength values required are:

σ_1^u = longitudinal strength (both tensile and compressive)

σ_2^u = transverse strength (both tensile and compressive)

τ_{12}^u = shear strength

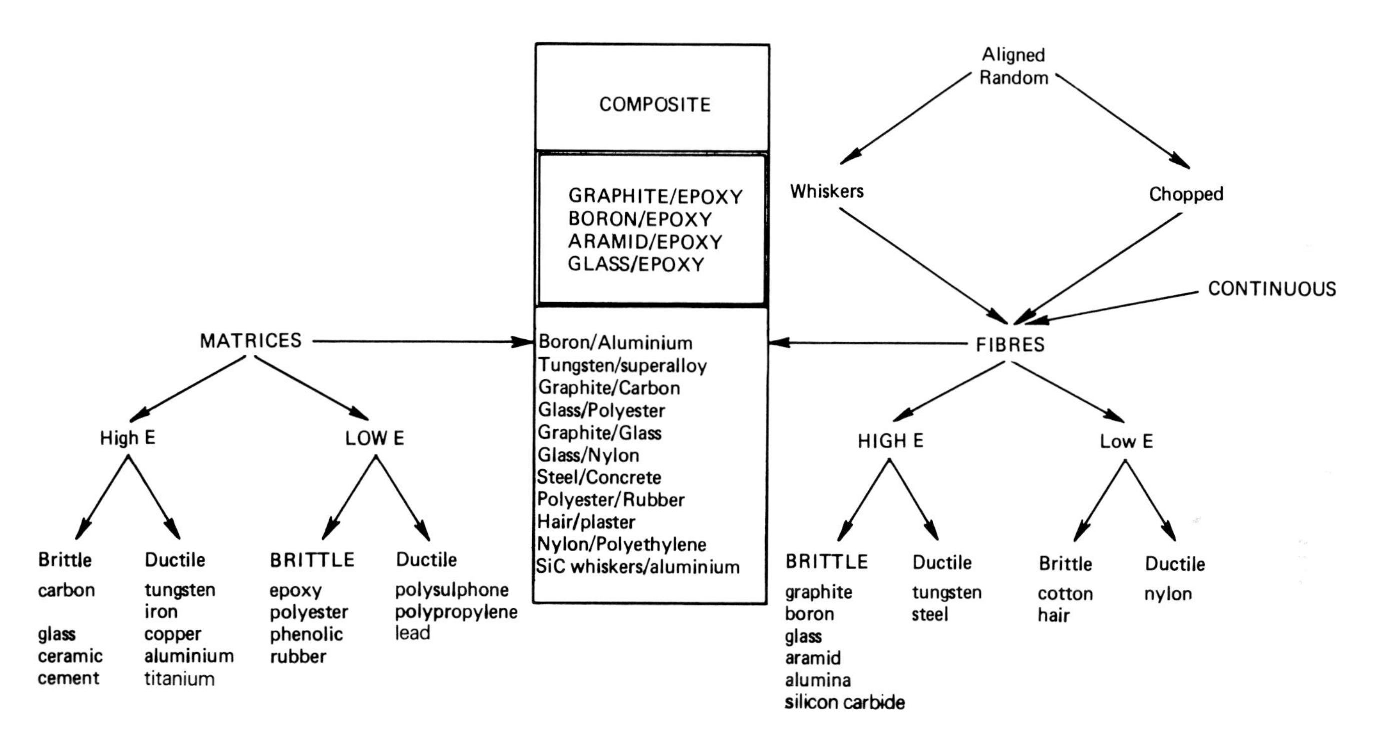

Fig. 2.1 Classification of composites according to fiber and matrix properties.

2.2 THE ELASTIC CONSTANTS

Mechanics of Materials Approach

The simple model used in the following analyses is a single, unidirectional ply, or "lamina," as depicted in Fig. 2.2. Note that the representative volume element shown is the full thickness of the single ply and that the simplified "two-dimensional" element is used in the following analyses. The key assumptions used in connection with this model are indicated in Fig. 2.3.

E_1, longitudinal modulus. The representative volume element under an applied stress σ_1 is shown in Fig. 2.3a. The resultant strain ε_1 is assumed to be common to both the fiber and matrix. The stresses felt by the fiber, matrix, and composite are, respectively, σ_f, σ_m, and σ_1. Taking E_f and E_m

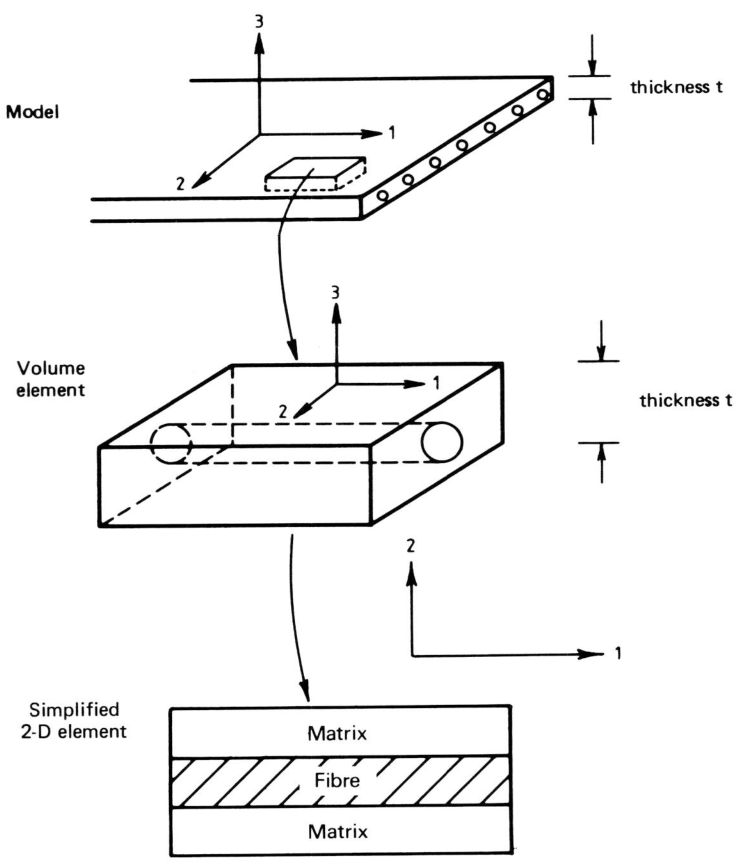

Fig. 2.2 Model and representative volume element of a single, unidirectional ply.

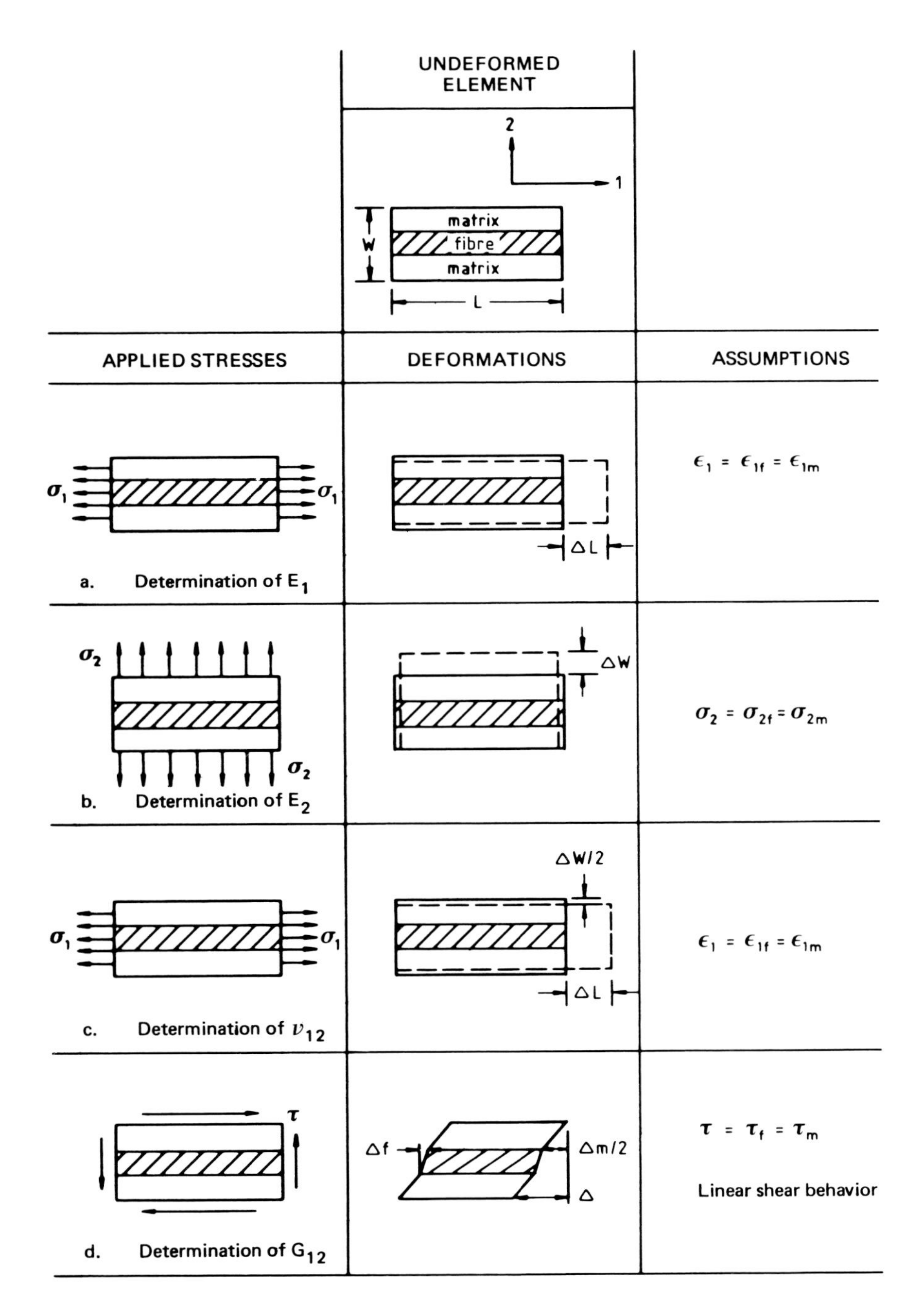

Fig. 2.3 Models for the determination of elastic constants by the "mechanics of materials approach."

as the fiber and matrix moduli, respectively, then

$$\sigma_f = E_f \varepsilon_1, \quad \sigma_m = E_m \varepsilon_1, \quad \sigma_1 = E_1 \varepsilon_1 \tag{2.1}$$

The applied stress acts over a cross-sectional area A consisting of A_f, the fiber cross section, and A_m, the matrix cross section. Since the fibers and matrix are acting in parallel to carry the load,

$$\sigma_1 A = \sigma_f A_f + \sigma_m A_m$$

or

$$\sigma_1 = \sigma_f V_f + \sigma_m V_m \tag{2.2}$$

where $V_f = A_f/A$ = fiber volume fraction and $V_m = A_m/A = 1 - V_f$ = matrix volume fraction.

Substituting Eq. (2.1) into Eq. (2.2) gives

$$E_1 = E_f V_f + E_m V_m \tag{2.3}$$

Equation (2.3) is a "rule of mixtures" type of relationship that relates the composite property to the weighted sum of the constituent properties. Experimental verification of Eq. (2.3) has been obtained for many fiber/resin systems; examples of the variation of E_1 with V_f for two glass/polyester resin systems are shown in Fig. 2.4.

E_2, transverse modulus. As shown in Fig. 2.3b, the fiber and matrix are assumed to act in series, both carrying the same applied stress σ_2. The transverse strains for the fiber, matrix, and composite are thus, respectively,

$$\varepsilon_f = \sigma_2/E_f, \quad \varepsilon_m = \sigma_2/E_m, \quad \varepsilon_2 = \sigma_2/E_2 \tag{2.4}$$

Deformations are additive over the width W, so that

$$\Delta W = \Delta W_f + \Delta W_m$$

or

$$\varepsilon_2 W = \varepsilon_f (V_f W) + \varepsilon_m (V_m W) \tag{2.5}$$

Substitution of Eq. (2.4) into Eq. (2.5) yields

$$1/E_2 = V_f/E_f + V_m/E_m \tag{2.6}$$

Experimental results are in reasonable agreement with Eq. (2.6) as shown, for example, in Fig. 2.5.

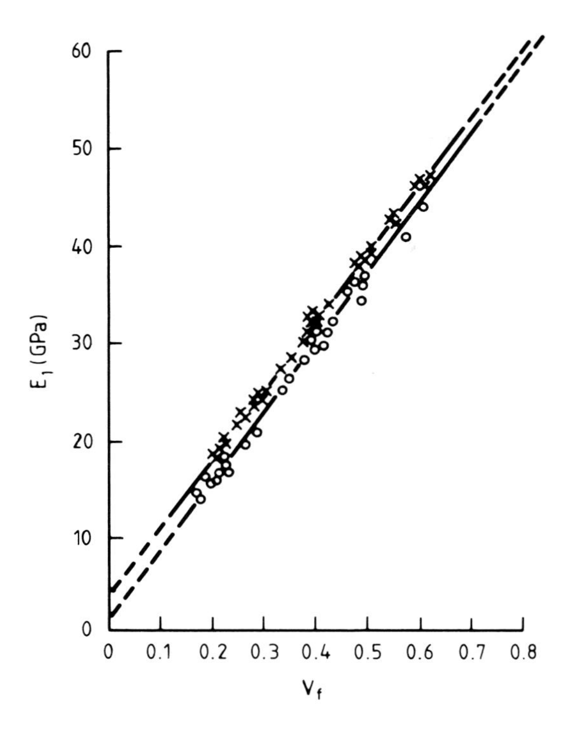

Fig. 2.4 E_1 **vs** V_f **for two glass/polyester systems.**

Several interesting features emerge from Eqs. (2.3) and (2.6). In high-performance composites, the fiber moduli are much greater than the resin moduli, so that, in the typical fiber volume fraction range of 50–60%, the matrix has only a small effect upon E_1 while the fibers have only a small effect on E_2. In other words,

$$E_1 \approx E_f V_f, \quad E_2 \approx E_m / V_m$$

ν_{12}, ***major Poisson's ratio.*** The major Poisson's ratio is defined by

$$\nu_{12} = -\varepsilon_2/\varepsilon_1 \tag{2.7}$$

where the only applied stress is σ_1 (Fig. 2.3c).

The transverse deformation is given by

$$\Delta W = \Delta W_f + \Delta W_m$$

or

$$\varepsilon_2 W = -\nu_f \varepsilon_1 (V_f W) - \nu_m \varepsilon_1 (V_m W) \tag{2.8}$$

since the fibers and matrix have equal strains in the longitudinal direction and, in general,

$$\nu = -\varepsilon_2/\varepsilon_1$$

Substituting for ε_2 from Eq. (2.7) into Eq. (2.8) gives the result

$$\nu_{12} = \nu_f V_f + \nu_m V_m \tag{2.9}$$

which is another "rule of mixtures" expression.

G_{12}, in-plane shear modulus. The applied shear stresses and resultant deformations of the representative volume element are shown in Fig. 2.3d. The shear stresses felt by the fiber and matrix are assumed equal and the composite is assumed to behave linearly in shear (which is, in fact, not true for many systems).

The total shear deformation is given by

$$\Delta = \gamma W$$

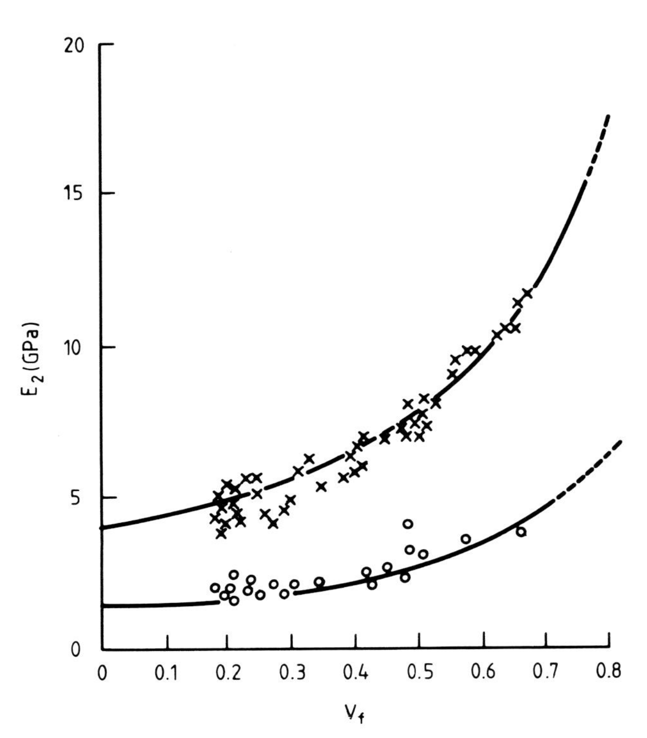

Fig. 2.5 E_2 vs V_f for two glass/polyester systems.

where γ is the shear strain of the composite. The deformation Δ consists of two additive components, so that

$$\Delta = \Delta_f + \Delta_m$$

or

$$\gamma W = \gamma_f(V_f W) + \gamma_m(V_m W) \tag{2.10}$$

Since equal shear stresses are assumed,

$$\gamma_f = \tau/G_f, \quad \gamma_m = \tau/G_m, \quad \gamma = \tau/G_{12} \tag{2.11}$$

substitution of Eq. (2.11) into Eq. (2.10) yields

$$1/G_{12} = V_f/G_f + V_m/G_m \tag{2.12}$$

Since G_m is very much smaller than G_f, the value of G_m has the major effect on G_{12} for typical 50–60% V_f values; the situation is analogous to that for the transverse modulus E_2.

Refinements to Mechanics of Materials Approach for E_1 and E_2

Prediction of E_1. Equation (2.3) is considered to provide a good estimate of the longitudinal modulus E_1. However, it does not allow for the triaxial stress condition in the matrix resulting from the constraint caused by the fibers. Ekvall[1] has produced a modified version of the equation to

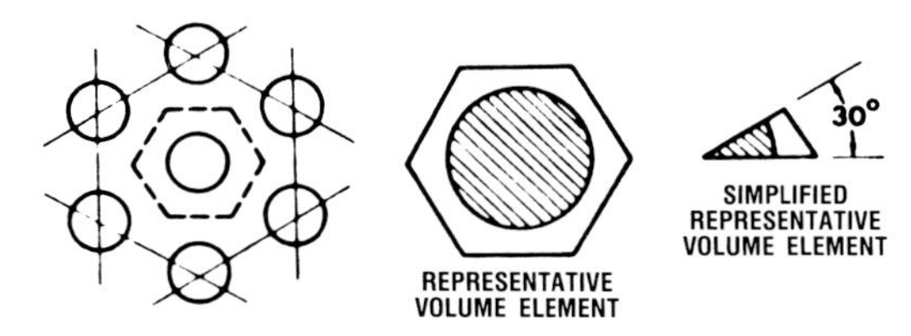

Hexagonal array and representative volume elements

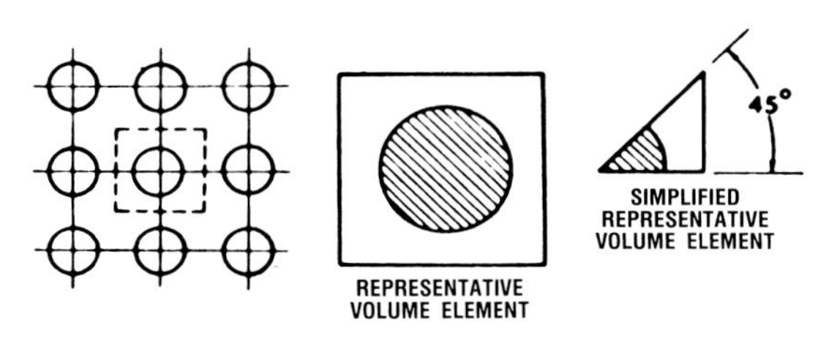

Square array and representative volume elements

Fig. 2.6 Typical models of composites for exact elasticity solutions.

allow for this effect,

$$E_1 = E_f V_f + E'_m V_m \tag{2.13}$$

where

$$E'_m = E_m/(1 - 2\nu_m^2)$$

and ν_m is Poisson's ratio for the matrix material. However, the modification is not large for values of ν_m of approximately 0.3.

Prediction of E_2. Equation (2.6) is considered to provide only an approximate estimate of the transverse modulus E_2. This is because, for loading in the transverse direction, biaxial effects resulting from differences in contraction in the longitudinal (fiber) direction between the fiber and the matrix become significant. The contraction difference arises because the two phases experience different strains and is even more marked if there is a difference in their Poisson's ratios.

The modified version of Eq. (2.6) produced by Ekvall[1] is

$$\frac{1}{E_2} = \frac{V_f}{E_f} + \frac{V_m}{E_m} - \frac{V_f}{E_f}\frac{[(E_f\nu_m/E_m) - \nu_f]^2}{[(V_f E_f/V_m E_m) + 1]} \tag{2.14}$$

Theory of Elasticity Approach to the Elastic Constants

The theory of elasticity approach to the determination of the elastic constants for composites is based on a wide variety of models and energy balance treatments. A detailed discussion of these approaches is beyond the scope of this chapter; however, some aspects are outlined here.

Energy approach. Bounding (or variational) derivations use energy balance considerations to produce upper and lower bounds on the elastic constants. The usefulness of the results, of course, depends upon the closeness of the bounds, as demonstrated in the following example.

Considering the stressed element shown in Fig. 2.3a, it can be shown[2] that the lower bound on the longitudinal modulus E_1 is given by

$$1/E_1 \leq V_m/E_m + V_f/E_f \tag{2.15}$$

[compare with Eq. (2.6)], while the upper bound is given by

$$E_1 \leq \frac{1 - \nu_f - 4\nu_f\nu_{12} + 2\nu_{12}^2}{1 - \nu_f - 2\nu_f^2} E_f V_f + \frac{1 - \nu_m - 4\nu_m\nu_{12} + 2\nu_{12}^2}{1 - \nu_m - 2\nu_m^2} E_m V_m \tag{2.16}$$

where

$$\nu_{12} = \frac{(1 - \nu_m - 2\nu_m^2)\nu_f E_f V_f + (1 - \nu_f - 2\nu_f^2)\nu_m E_m V_m}{(1 - \nu_m - 2\nu_m^2) E_f V_f + (1 - \nu_f - 2\nu_f^2) E_m V_m}$$

It is of interest to note that if $\nu_{12} = \nu_f = \nu_m$, the upper bound solution becomes

$$E_1 \leq E_f V_f + E_m V_m$$

the same result as Eq. (2.3), which implies an equality of ν_f and ν_m in the mechanics of materials approach.

In this example, the bounding solutions are not very useful because the bounds are too far apart, the lower bound being the transverse modulus as predicted by the mechanics of materials approach.

Direct approaches. Here, various representative models of elastic inclusions in an elastic matrix are employed to obtain exact solutions for the stiffness properties. Typical volume elements assumed for a hexagonal and a square fiber distribution are shown in Fig. 2.6. In many cases the solutions are highly complex and of limited practical use. Regular fiber distributions do not occur in practical composites. Rather, the array is random and the analysis for regular arrays must be modified to allow for the extent of contact between fibers; this is called the "degree of contiguity"[3] and is measured by a coefficient c, which can vary from $c = 0$ for isolated fibers to $c = 1$ for contacting fibers. This situation is illustrated in Fig. 2.7. The effective value of c may be determined experimentally. The degree of contiguity has more effect on E_2 and G_{12} than on E_1. These matters, and other simplifying approaches, such as the Halpin-Tsai equations, are discussed more fully in Ref. 4.

2.3 MECHANICS OF MATERIALS APPROACH TO STRENGTH

Simple Estimate of Tensile Strength

The simplest analysis of longitudinal tensile strength assumes that all fibers break at the same stress level, at the same time, and in the same plane. Although this assumption is grossly unrealistic, it provides a starting point for further analysis.

As with the model used to determine E_1, the fibers and matrix are assumed to experience equal strains. In advanced epoxy/matrix composites, the strain-to-failure capability of the stiff fibers, ε_f^u, is markedly less than that of the matrix, ε_m^u, as shown in Fig. 2.8a. The fibers will thus fail first and the total load will be transferred to the matrix. Two composite failure modes can be envisaged depending on the fiber volume fraction V_f. At high V_f, the matrix alone is not capable of bearing the full load and fractures immediately after fiber fracture. The composite strength is thus given by

$$\sigma_1^u = \sigma_f^u V_f + \sigma_m' V_m \tag{2.17}$$

where σ_f^u is the fiber failure stress and σ_m' is defined in Fig. 2.8a as the stress carried by the matrix material at the fiber breaking strain. At low V_f, there is enough matrix material to carry the full load after the fibers fracture; the

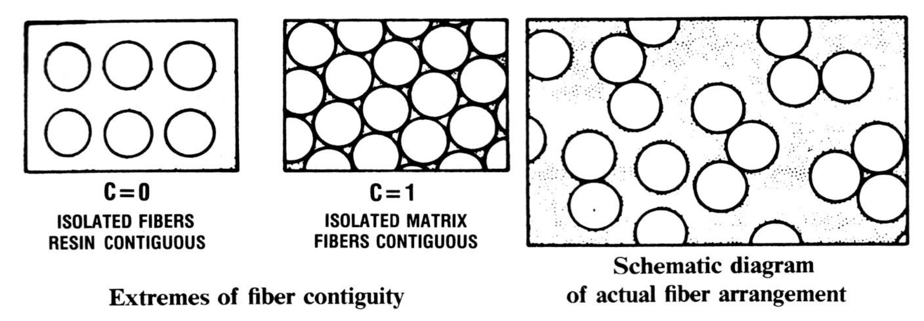

Fig. 2.7 Concept of contiguity used for semiempirical elasticity solutions.

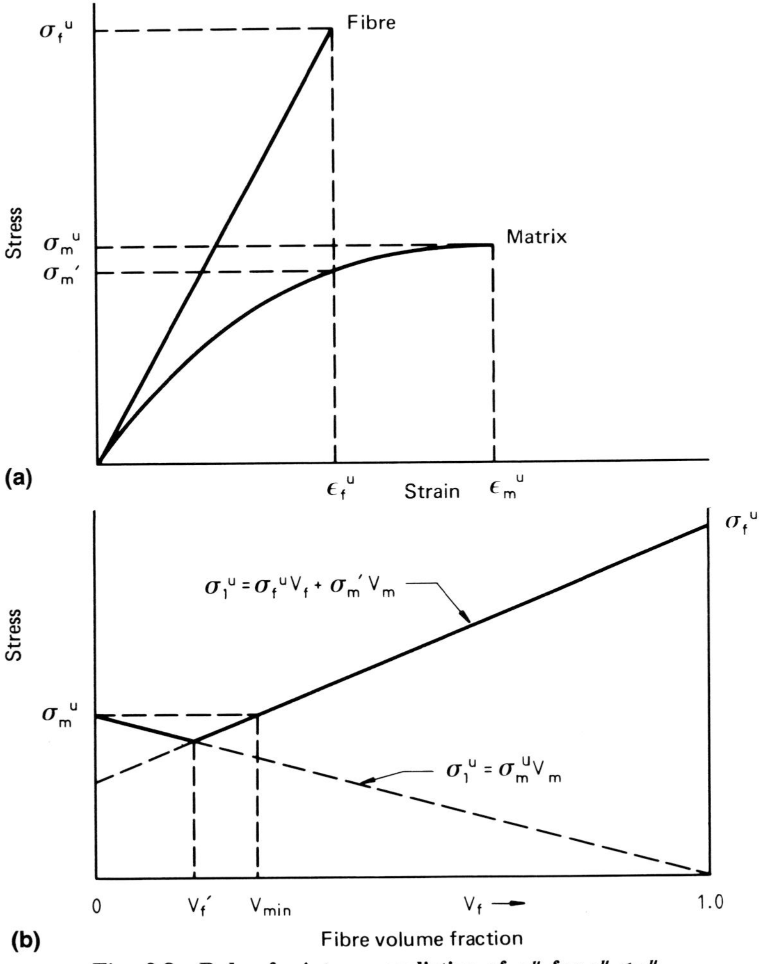

Fig. 2.8 Rule of mixtures prediction of σ_1^u for $\varepsilon_f^u < \varepsilon_m^u$.

composite strength is then given by

$$\sigma_1^u = \sigma_m^u V_m$$

σ_1^u is plotted as a function of V_f in Fig. 2.8b, where it can be readily seen that the value of V_f corresponding to a change in the failure mode is given by

$$V_f' = (\sigma_m^u - \sigma_m')/(\sigma_f^u + \sigma_m^u - \sigma_m') \tag{2.18}$$

Note also that there is a minimum volume fraction $V_{\min}$ below which composite strength is actually less than the inherent matrix strength,

$$V_{\min} = (\sigma_m^u - \sigma_m')/(\sigma_f^u - \sigma_m') \tag{2.19}$$

For high-strength, high-modulus fibers in relatively weak, low-modulus epoxy matrices, σ_m', V_f', and $V_{\min}$ will be quite small.

Analogous treatments can be applied to systems in which the matrix fails first, but obviously the physical characteristics of the fracture modes will be quite different.

Statistical Analysis of Tensile Strength

General. The foregoing analysis of tensile strength assumed simultaneous fracture of equal-strength fibers in one plane. In reality, the situation

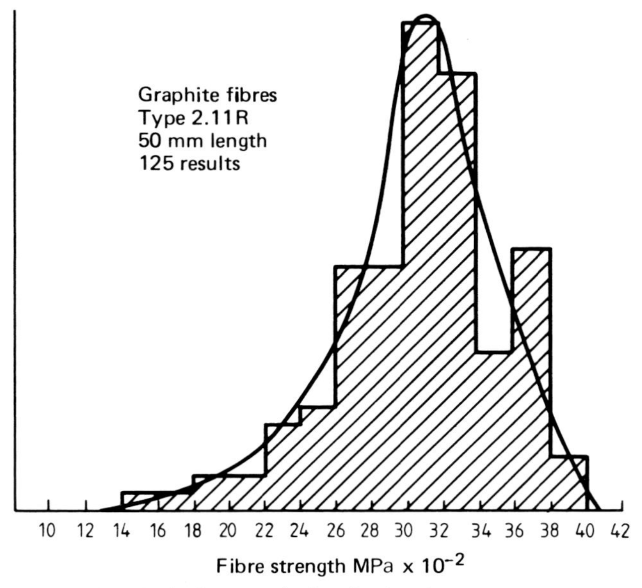

Fig. 2.9 Typical strength distribution for graphite fibers.

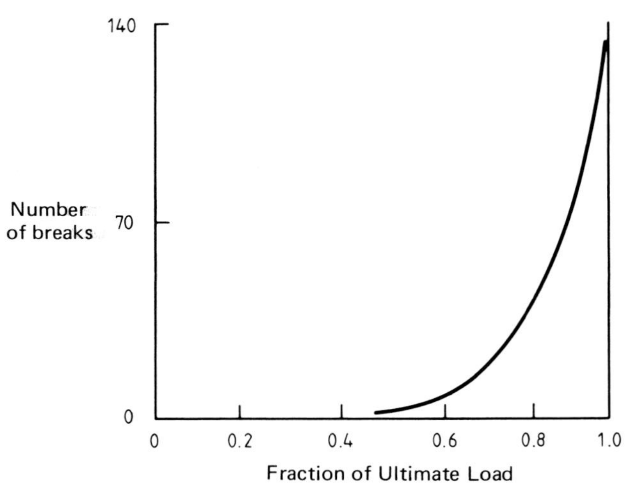

Fig. 2.10 Number of fiber fractures vs fraction of ultimate composite strength.

is much more complex, because of the brittle (flaw-sensitive) nature of the fibers and the fiber/matrix interaction. These two features are discussed below.

Brittle fibers contain surface flaws or imperfections that produce "weak spots" along the fiber length. Fiber fractures will occur at these flaws at more or less random positions throughout the composite. Therefore, fracture will not occur in a single plane. In the most simple case, in which the imperfect fibers all have the same strength and the matrix is unable to grip the broken fibers, the strength of the composite would be as calculated in the previous subsection.

For brittle fibers, however, flaws vary not only in position, but also in severity. The way in which the fiber strength changes as a result of this variation in the flaw severity is shown in Fig. 2.9 for a typical case. Therefore, it would be expected that fiber fractures would occur throughout a range of stress levels, up to ultimate composite failure. This is indeed the case, as shown in Fig. 2.10.

Another important characteristic of composite fracture is the fiber/matrix interaction in the vicinity of a fiber fracture. Rather than becoming ineffective, a broken fiber can still contribute to composite strength because the matrix is able to transfer stress back into the fiber from the broken end, as shown in Fig. 2.11. High shear stresses develop in the matrix and then decay a short distance from the break; at the same time, the tensile stress carried by the fiber increases from zero at the broken end to the full stress carried by unbroken fibers. The characteristic length over which this stress buildup occurs is known as the ineffective length δ (see Fig. 2.11).†

†Often, the term "critical transfer length" is used in this context, the critical transfer length being twice the ineffective length.

The ineffective length can be determined experimentally by measuring the stress required to pull fibers of various lengths from a matrix.[5]

If the fiber/matrix bond strength is low, the high shear stresses will cause fiber/matrix debonding, as shown in Fig. 2.12a. It is also possible that the stress elevation felt by fibers adjacent to the fractured fiber (Fig. 2.11) is sufficient to cause further fiber fractures and crack propagation through the

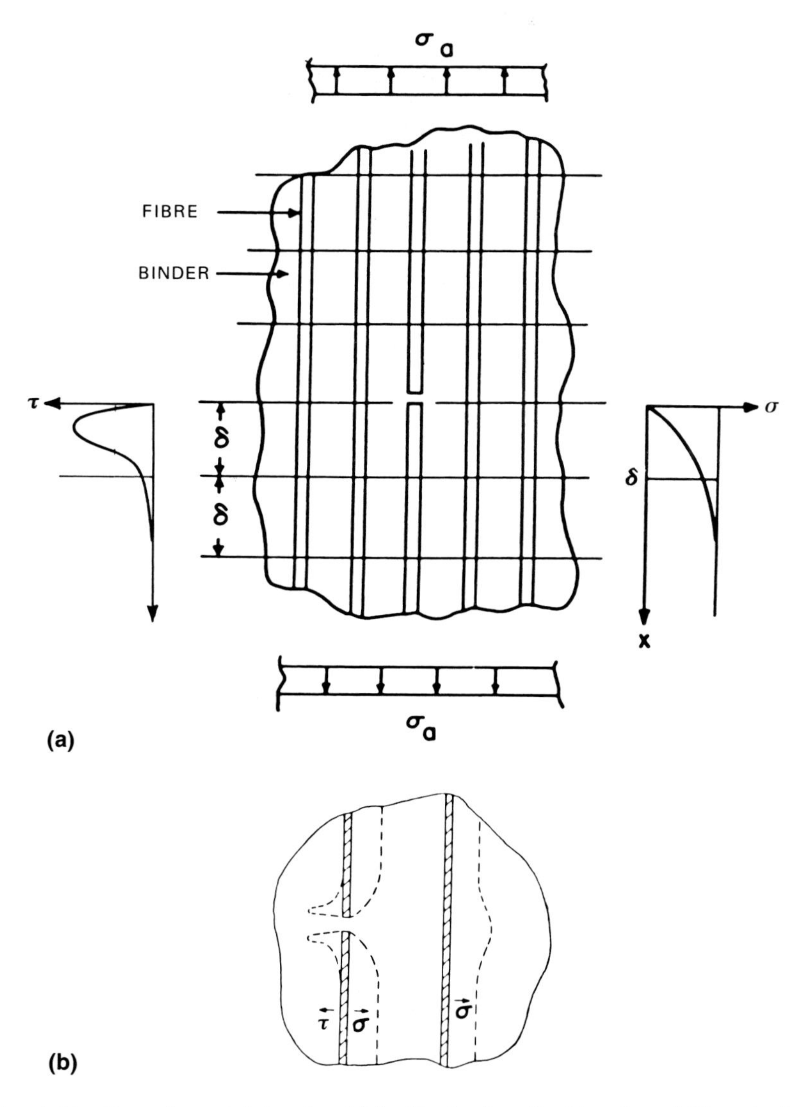

Fig. 2.11 Rosen's model showing (a) ineffective length at break and (b) perturbation of stress in adjacent fiber.

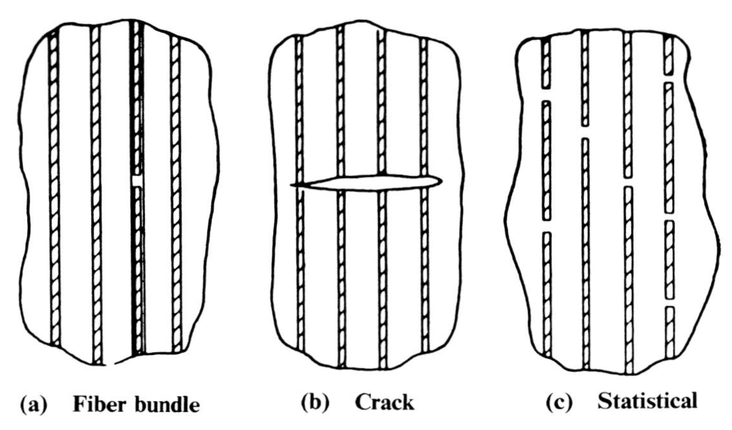

(a) Fiber bundle (b) Crack (c) Statistical

Fig. 2.12 Possible composite tensile failure modes.

matrix in the brittle fashion shown in Fig. 2.12b. In addition to the stress concentration felt by fibers resulting from the ineffectiveness of adjacent fractured fibers near the broken ends, there is also a stress concentration associated with the crack in the matrix surrounding the fiber fracture.

If fiber/matrix debonding, crack propagation, and fiber fracture propagation are not the dominant failure mechanisms, the composite will accumulate damage under increasing stress and failure will occur by the cumulative damage mechanism outlined below and shown in Fig. 2.12c.

Rosen's model of cumulative damage. The strength of individual fibers is dependent on the probability of finding a flaw and, therefore, on the fiber length. It has been shown that the strength/length relationship takes the form of a Weibull distribution of the form

$$f(\sigma) = L\alpha\beta\sigma^{\beta-1}\exp(-L\alpha\sigma^{\beta}) \tag{2.20}$$

where $f(\sigma)$ is the probability density function for fiber strength σ, L is the fiber length, and α and β are the material constants.

Constant α determines the position of the Weibull distribution, while constant β determines its shape. Both α and β are experimentally accessible quantities and can be determined, for example, from a log-log plot of mean fiber strength for fibers of given lengths vs fiber length.

Daniels[6] showed that the strength of a bundle of N fibers having such a Weibull distribution can be described by a normal distribution where the mean value $\bar{\sigma}_{BL}$ is a function of fiber length,

$$\bar{\sigma}_{BL} = (L\alpha\beta)^{-1/\beta}\exp(-1/\beta) \tag{2.21}$$

and whose standard deviation is proportional to $N^{-\frac{1}{2}}$. Thus, for very large N, all of the bundles tend toward the same strength value $\bar{\sigma}_{BL}$.

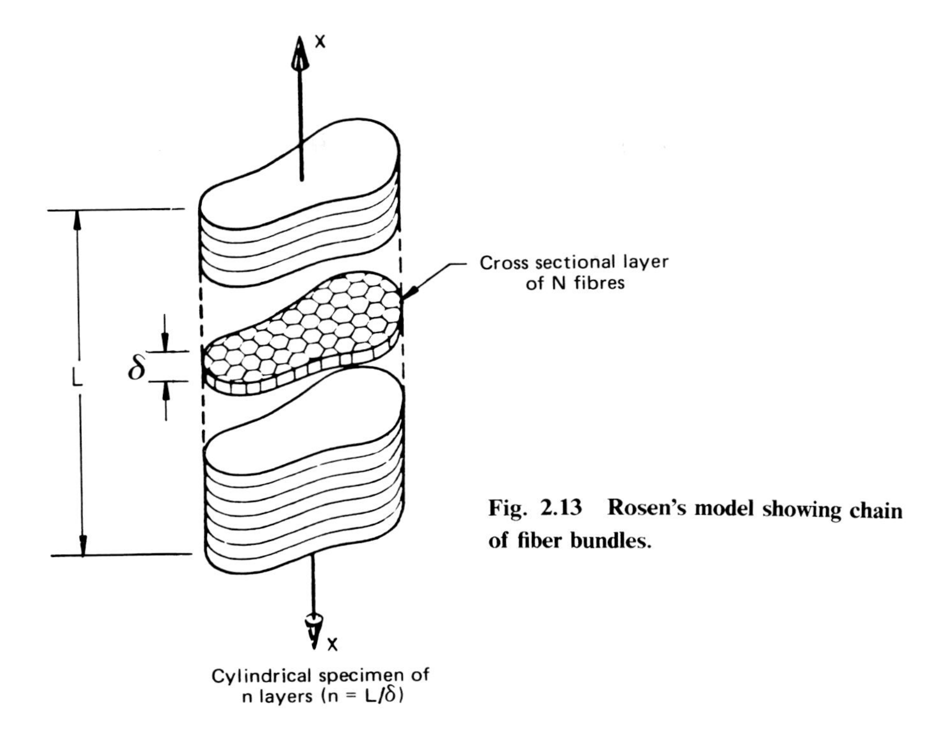

Fig. 2.13 Rosen's model showing chain of fiber bundles.

Rosen[7] models the composite as a chain of bundles (Fig. 2.13), the length of each bundle (or chain link) being the ineffective length δ. For very large N, the strength of each bundle or chain link will be the same and the strength of the whole chain (or composite) will be equal to the link strength which is given by

$$\bar{\sigma}_{B\delta} = (\delta\alpha\beta)^{-1/\beta}\exp(-1/\beta) \tag{2.22}$$

Thus, it is possible to compare the strengths of a bundle of "dry" fibers of length L and a composite with ineffective length δ as follows:

$$\bar{\sigma}_{B\delta}/\bar{\sigma}_{\mathrm{BL}} = (L/\delta)^{1/\beta} \tag{2.23}$$

For graphite fibers in an epoxy matrix, $\beta \cong 10$ and $\delta \cong 10^{-2}$ mm (about a fiber diameter), so if $L = 100$ mm, then

$$\bar{\sigma}_{B\delta}/\bar{\sigma}_{\mathrm{BL}} = (100/10^{-2})^{1/10} \cong 2.5$$

This is the strengthening obtained by composite action.

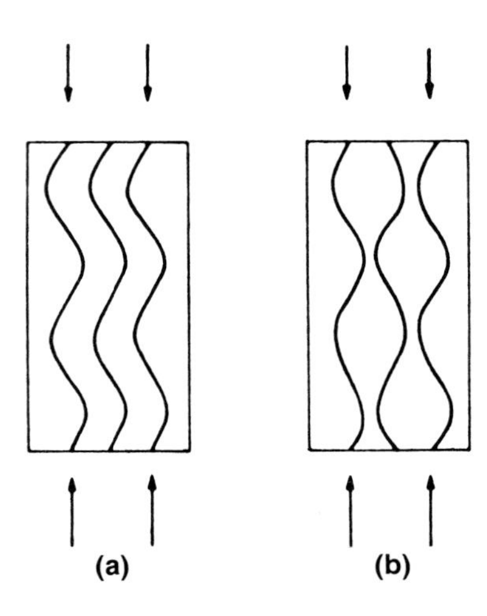

Fig. 2.14 Two pure buckling modes for unidirectional composites in compression: (a) shear or in-phase mode and (b) extensional or out-of-phase mode.

Mechanics of Materials Approach to Compressive Strength

The previously introduced equations relating to tensile failure do not apply to compressive strength because fibers do not fail in simple compression. Rather, they fail by local buckling. This behavior is very complicated since it is dependent on the presence of residual stresses in the matrix caused by different fiber and matrix expansion coefficients. It has been shown, for example, that glass fibers in an epoxy matrix will buckle after cool down from the resin cure temperature. As may be expected, by assuming that the fibers act as circular columns on an elastic foundation, the wavelength of the buckling increases with the fiber diameter.

Two pure buckling modes can be envisaged (see Fig. 2.14): (1) the extensional mode, in which the matrix is stretched and compressed in an out-of-phase manner or (2) the shear mode in which the fibers buckle in phase and the matrix is sheared. The most likely mode is that producing the lowest energy in the system. While mixed modes are possible, they require more energy than either of the pure modes.

Analysis of the buckling[4] is based on the energy method in which the change in strain energy of the fibers ΔU_f and of the matrix ΔU_m, as the composite changes from the compressed but unbuckled state to the buckled state, is equated to the work done ΔW by the external loads

$$\Delta U_f + \Delta U_m = \Delta W$$

In the model, the composite is considered two-dimensional, the fibers are treated as plates normal to the plane of Fig. 2.15 (rather than rods) and the buckling pattern is assumed to be sinusoidal. The resulting buckling stress for the extensional mode is

$$\sigma_{c\,\max} \cong 2V_f[(V_f E_m E_f)/3(1 - V_f)]^{\frac{1}{2}} \tag{2.24}$$

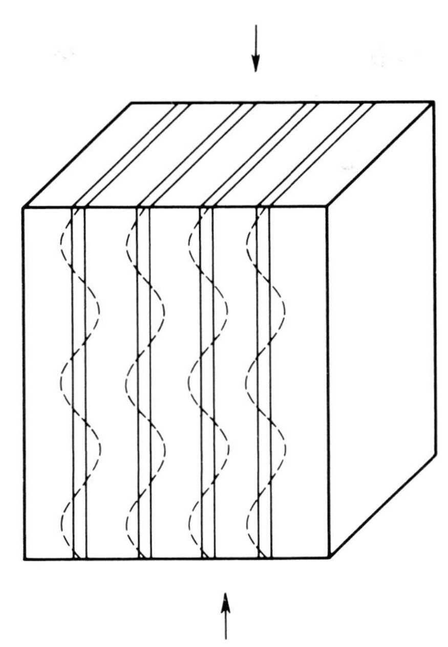

Fig. 2.15 Model for buckling fibers in a unidirectional composite under compression (fibers are considered plates rather than rods; buckling is assumed to be sinusoidal).

and for the shear mode

$$\sigma_{c\,\max} = G_m/(1 - V_f) \tag{2.25}$$

As V_f tends to zero, $\sigma_{c\,\max}$ for the extensional mode tends to zero; but, as V_f tends to unity, $\sigma_{c\,\max}$ for the extensional mode becomes very large compared with $\sigma_{c\,\max}$ for the shear mode. Thus, the extensional mode would be expected to apply for only small V_f. From Ref. 5, assuming $E_f \gg E_m$ and $\nu_m = \frac{1}{3}$, with $G_m = E_m/2(1 + \nu_m)$, then the transition occurs at $(E_m/10E_f)^{\frac{1}{3}}$, or at $V_f = 10\%$ for $E_f/E_m = 100$, and at $V_f = 22\%$ for $E_f/E_m = 10$. It has been found that these equations tend to overpredict considerably the compressive strength.

One exception is boron/epoxy, whose actual compressive strength is only about 63% lower than that predicted. Generally, the problem is that the predicted failure strains are much higher than the matrix yield strain (e.g., 5% in the case of glass/epoxy). As an approximation to the inelastic behavior, the theory was expanded in Ref. 8 using a gradually reducing matrix shear modulus. This gives more reasonable agreement with experimental results.

A very simple approach that appears to predict the experimental behavior in some cases is obtained by assuming failure occurs when the matrix

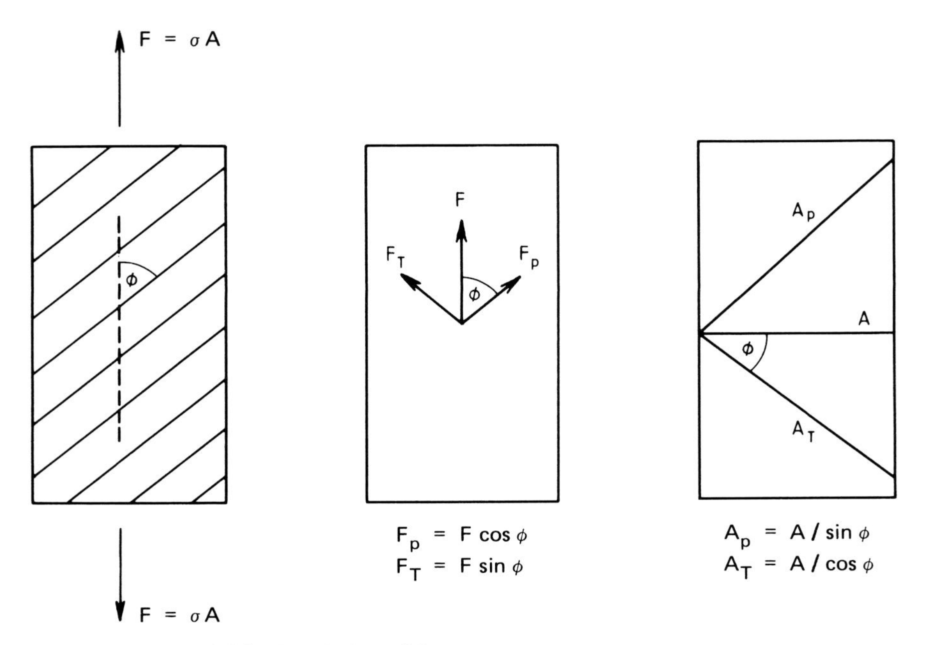

Fig. 2.16 Resolution of forces and area in off-axis tension.

reaches its yield stress σ_m^y. Thus, at failure

$$\sigma_{c\,\max} = \sigma_f V_f + \sigma_m^y V_m \tag{2.26}$$

where the fiber stress σ_f is given by strain compatibility as

$$\sigma_f = \sigma_m^y E_f / E_m = \varepsilon_m^y E_f \tag{2.27}$$

Taking ε_m^y for an epoxy resin as 0.02 and $E_f = 70$ GPa for glass fibers, then

$$\sigma_f = 0.02 \times 70 \text{ GPa} = 1.4 \text{ GPa}$$

and so

$$\sigma_c^u = \sigma_f V_f = 1.4 \times 0.6 = 0.84 \text{ GPa} \quad \text{(ignoring the small contribution from the matrix)}$$

This result is in reasonable agreement with typically observed values.

Off-Axis Strength in Tension

The failure of an oriented, but still unidirectional composite can be envisaged as occurring in any of three modes: (1) failure normal to the fibers

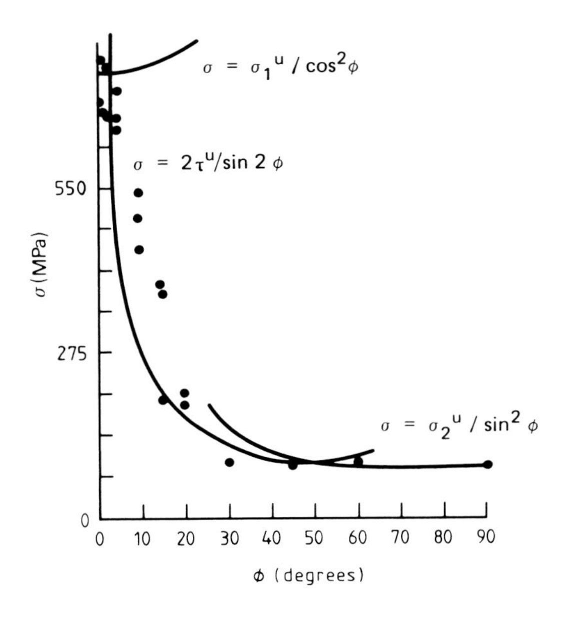

Fig. 2.17 Example of the variation of tensile strength vs orientation for unidirectional composites.

(as occurs with straight tension in the fiber direction), (2) failure parallel to the fibers by matrix rupture or fiber/matrix interface tensile failure, and (3) failure by shear of the matrix or fiber/matrix interface.

If the fibers make an angle ϕ with the direction of applied tensile stress σ, then, as shown in Fig. 2.16, the stresses can be resolved as follows:

Tensile stress parallel to fibers: $\sigma_1 = \sigma \cos^2\phi$

Tensile stress normal to fibers: $\sigma_2 = \sigma \sin^2\phi$

Shear stress parallel to fibers: $\tau_{12} = \frac{1}{2}\sigma \sin 2\phi$

If σ_1^u, σ_2^u, and τ^u represent the composite strengths in direct tension ($\phi = 0°$), transverse tension ($\phi = 90°$), and shear ($\phi = 45°$), respectively, then the failure stress σ for each mode can be expressed as

$$\text{Mode 1: } \sigma = \sigma_1^u/\cos^2\phi$$

$$\text{Mode 2: } \sigma = \sigma_2^u/\sin^2\phi$$

$$\text{Mode 3: } \sigma = 2\tau^u/\sin 2\phi \qquad (2.28)$$

Thus, the failure mode changes with ϕ as shown in Fig. 2.17. Although these results are obeyed quite well for many systems and the observed fracture modes are as predicted, the interaction of stresses and the occurrence of mixed-mode fractures are not accounted for. Reference 4 presents a more detailed analysis that accounts for the complex stress states.

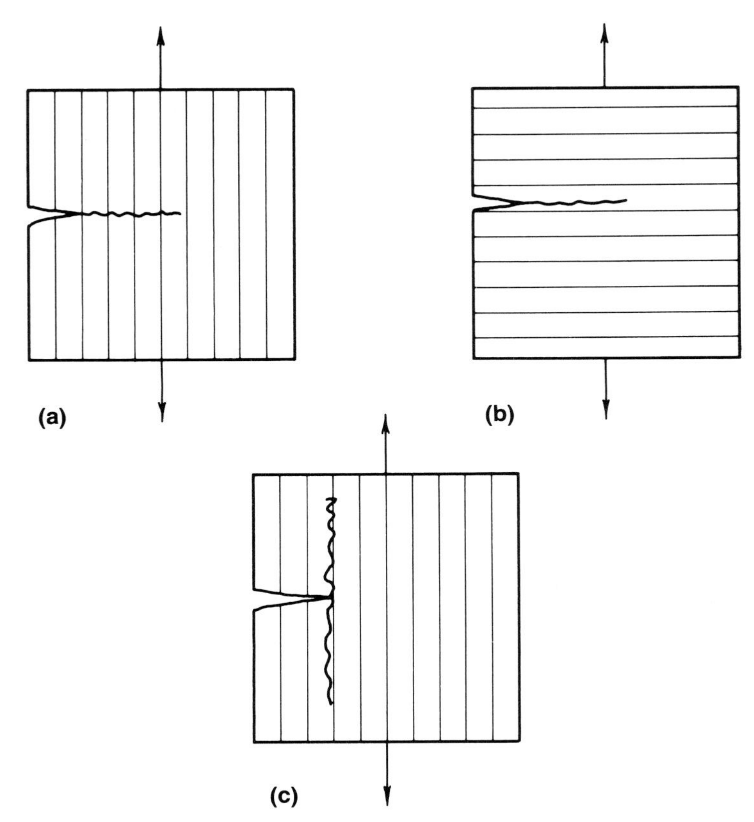

Fig. 2.18 Three basic modes of crack propagation in unidirectional fiber composites subjected to simple tensile loading: (a) normal to the fibers, (b) parallel to the fibers, and (c) splitting. Modes a and b are self-similar modes of propagation.

Figure 2.17 shows that strength falls rapidly with increasing ϕ. However, if the plies are placed at $+\phi$ and $-\phi$, the rate of fall-off is very much less, even to values of ϕ as high as 30°.

2.4 FRACTURE TOUGHNESS

Fracture Surface Energy

A measure of the toughness, or the resistance of a material to crack propagation, is its fracture surface energy γ. This is defined as the minimum amount of energy required to create the unit area of free surface and is usually given in units of kJ/m^2. Since two free surfaces are produced, R (for crack resistance) equal to 2γ is the term often employed in fracture calculations.

It is a matter of considerable importance that, for crack propagation normal to the fibers (Fig. 2.18a), the fracture energy of a composite

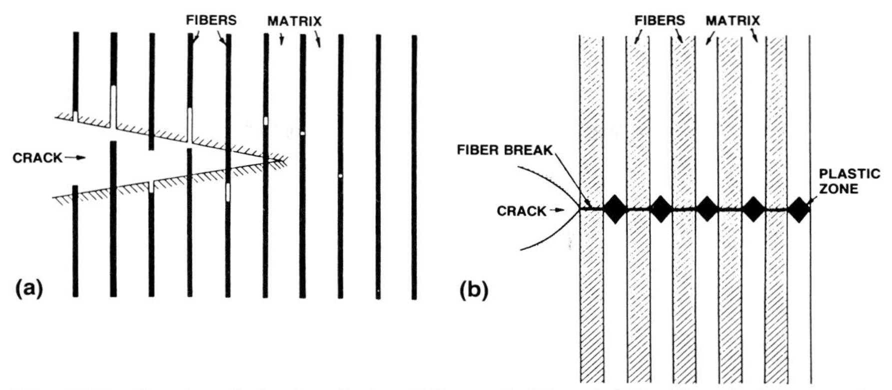

Fig. 2.19 Fracture behavior during failure of (a) a brittle fiber/brittle matrix composite and (b) a brittle fiber/ductile matrix composite. In (b), the fibers are assumed to be strongly bonded to the matrix. (Taken from Ref. 9.)

consisting of brittle fibers in a brittle matrix is usually very much greater than is predicted by a simple rule of mixtures relationship. In general, $R_1 \gg V_f R_f + V_m R_m$. For example, in the case of a typical graphite/epoxy composite, $R_m \approx 1$ kJ/m^2 for the bulk epoxy resin and $R_f \approx 0.1$ kJ/m^2 for the graphite material. However, the fracture surface energy of a unidirectional composite R_C (if the crack is made to propagate normal to the fibers) is typically 25–50 kJ/m^2.

In contrast, for crack propagation parallel to the fibers (Fig. 2.18b), the fracture surface energy R_2 is of the order of R_m if the crack propagates solely through the matrix; however, R_m may be lower if the crack propagates, partially, through the weaker fiber/matrix interface. Because $R_2 \ll R_1$, crack propagation parallel to the fibers, or splitting (Fig. 2.18c), may occur even when the starting crack is normal to the fibers.

Considering crack propagation normal to the fibers (Fig. 2.18a), the total work of fracture can be attributed to a number of sources, as shown in Fig. 2.19 and Table 2.1 taken from Ref. 9. In the case of the brittle fiber/brittle matrix composite (Fig. 2.19a), crack growth proceeds by pulling fibers out of the matrix behind the crack front and by fracturing fibers ahead of the crack tip. Energy is absorbed during *pull-out*, if the shear stress at the fiber-matrix interface is maintained while the fracture surface is separating. If the fiber/matrix interface is relatively weak, local stresses will cause the fibers to be *debonded* from the matrix, with a resultant loss of stored strain energy. Stored strain energy is also lost by *stress relaxation* over the transfer length when the fiber fractures. Finally, strain energy is also lost from the fiber by *crack bridging* if the fiber spans the opening crack prior to fracture.

If the matrix is ductile, as in a metal matrix composite, energy is also absorbed by *matrix plastic deformation*. This situation is illustrated in Fig. 2.19b for the case where the fiber is very strongly bonded to the matrix so that little fiber pull-out occurs.

The work of the fracture contribution for pull-out in Table 2.1 refers to short constant-strength fibers. The expression would have to be modified for a statistical flaw distribution in continuous fibers. The contribution due to creation of new surface of fibers and matrix can be ignored in brittle systems, as can contributions from the matrix yield.

Fracture Mechanics

The energetic requirement for crack propagation is that the energy release rate (equal to the "fixed grips" strain energy release rate) G must equal the fracture surface energy R. At the critical condition,

$$G = G_c = R \tag{2.29}$$

In many cases, it is more convenient to work in terms of the stress intensity factor K. For an isotropic material in the crack opening mode, K_I is related to G through the equation: $K_I^2 = EG$ and $K_{Ic}^2 = EG_c$ at failure.

In the simple case of a small center crack in a sheet under tension,

$$K_I = \sigma(\pi a)^{\frac{1}{2}} \tag{2.30}$$

where σ is the applied stress and $2a$ the crack length. Thus, the stress σ_F at failure is given by

$$\sigma_F = (ER/\pi a)^{\frac{1}{2}} \tag{2.31}$$

which is the familiar Griffith equation.

Table 2.1 Fracture Mechanisms[a]

Model	γ
Pull-out (short fibers of length L)	$\dfrac{V_f \sigma_f^u \delta^2}{3L} (L > 2\delta)$
	$\dfrac{V_f \sigma_f^u L^2}{24\delta} (L < 2\delta)$
Debonding	$\dfrac{V_f (\sigma_f^u)^2 y}{4E_f}$
Stress relaxation	$\leq \dfrac{V_f (\sigma_f^u)^2 \delta}{3E_f}$
Crack bridging	$\dfrac{2V_f r (\sigma_f^u)^3}{\tau_i E_f} \times \dfrac{(1-\nu_f)(1-2\nu_f)}{12(1-\nu_f)}$
Matrix plastic deformation	$\dfrac{(1-V_f)^2}{V_f} \times \dfrac{\sigma_m^u r}{\tau_m} \times U$

[a]Notation: y = debonded length of fiber, r = fiber radius, τ_i = shear strength of interface, τ_m = shear strength of matrix, U = work to fracture unit volume of matrix, δ = ineffective length $(= \frac{1}{2}\sigma_f^u r/\tau_i)$.

A relationship between G and K can also be obtained for a fiber composite material by modeling it as a continuous linear orthotropic material with the crack propagating on one of the planes of symmetry. Using the analysis for such materials given in Ref. 10,

$$G = K_I^2\left(\frac{a_{11}a_{22}}{2}\right)^{\frac{1}{2}}\left[\left(\frac{a_{22}}{a_{11}}\right)^{\frac{1}{2}} + \left(\frac{a_{66}+2a_{12}}{2a_{11}}\right)\right]^{\frac{1}{2}} \tag{2.32}$$

Here the a_{ij} are the coefficients of the stresses in the stress-strain law defined using a coordinate system based not on the fiber direction, but on the crack in the "1 direction" and the load in the "2 direction"; this seems to be conventional in fracture mechanics treatments of this type. The factor involving the a terms can be considered as the reciprocal of the effective modulus E' of the composite. For the simple case illustrated in Fig. 2.18a, where the crack is perpendicular to the fiber,

$$a_{22} = 1/E_{11}, \quad a_{11} = 1/E_{22}, \quad a_{66} = 1/G_{12}$$

$$a_{12} = -\nu_{12}/E_{11} = -\nu_{21}/E_{22}$$

Taking as values typical of a graphite/epoxy composite, $E_{11} = 140$ GPa, $E_{22} = 12$ GPa, $G_{12} = 6$ GPa, $\nu_{12} = 0.25$, and $\nu_{21} = 0.0213$, then $E' \approx 0.4E_{11}$. If, alternatively, the crack is considered to lie parallel to the fibers (Fig. 2.18b), then $E' \approx 2E_{22}$.

As long as crack propagation occurs on a plane of symmetry, the relationship between K_I, σ, and a for the linear orthotropic material remains the same as that for the isotropic material. Thus, referring back to the unidirectional composite, it can be seen that for a given crack length a and normal stress σ, a crack parallel to the fibers produces a larger G than one normal to the fibers. This results from the lower compliance of the composite when stressed in the fiber direction.

In the fiber composite material, the orientation dependence of R must also be taken into consideration. In general, a composite is notch sensitive to cracks running parallel to the fibers (Fig. 2.18b) and the fracture mechanics principles described above may be directly employed. However, the composite may not be notch sensitive in the situation shown in Fig. 2.18a. In some cases,[5] the composite may become notch sensitive when $a \gg \delta$, the ineffective length, since the strain concentration in the matrix may then lead to fiber fractures at the crack tip. The crack would then become more effective with increasing a, as required in fracture mechanics considerations. However, in other cases, gross failure of the fiber/matrix interface may occur, resulting in the splitting mode of failure illustrated in Fig. 2.18c. This situation, which occurs in more weakly bonded composites, results in complete notch insensitivity (failure at the net section strength). Both of these situations are illustrated in some experimental work on graphite/epoxy composites in Fig. 2.20, taken from Ref. 9.

The conditions for crack turning or splitting can be approached from energy considerations.[11] For the cracking illustrated in Fig. 2.18c to occur in

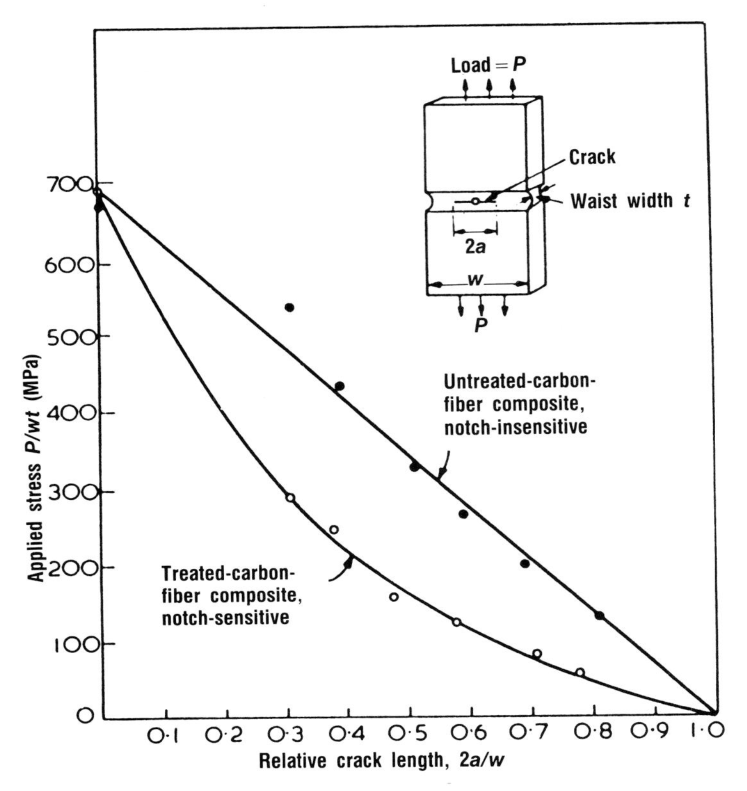

Fig. 2.20 Nominal applied stress vs relative crack size for a carbon fiber (graphite/epoxy) composite showing notch-insensitive and notch-sensitive behavior. (Taken from Ref. 9.)

a unidirectional composite, it is necessary that

$$G_1/G_2 < R_1/R_2$$

where G_1 is the energy release rate for self-similar propagation and G_2 the release rate for splitting. Typically, for graphite/epoxy R_1/R_2 is in the range 25–50 and G_1/G_2 is about 20, so splitting is generally predicted.

As mentioned earlier, most advanced composite structures are made of laminates in the form of unidirectional plies laminated together at various orientations. The fracture behavior of these materials is considered in Chap. 10. However, it is worth mentioning here that an approach similar to that just described can be taken for this highly complex situation.[12] The direction of crack growth is based on the R_1/R_2 ratio for each ply and the G_1/G_2 ratio for the laminate. The energy release rates are calculated by finite element procedures and recalculated after each increment of crack growth until the point of catastrophic failure.

References

[1]Ekvall, J. C., "Structural Behavior of Monofilament Composites," *AIAA 6th Structures and Materials Conference*, Palm Springs, CA, April 1965.

[2]Fung, Y. C., *Foundations of Solid Mechanics*, Prentice-Hall, Inc., Englewood Cliffs, NJ, 1965.

[3]Tsai, S. W., "Structural Behavior of Composite Materials," NASA CR-71, July 1964.

[4]Jones, R. M., *Mechanics of Composite Materials*, McGraw-Hill Book Co., New York, 1975.

[5]Kelly, A., *Strong Solids*, Oxford University Press, London, 1973.

[6]Daniels, H. E., "The Statistical Theory of the Strength of Bundles of Threads I," *Proceedings of the Royal Society of London*, Vol. A183, 1945, p. 405.

[7]Rosen, B. W., "Tensile Failure of Fibrous Composites," *AIAA Journal*, Vol. 2, Nov. 1964, pp. 1985-1991.

[8]Dow, Norris F. and Rosen, B. W., "Evaluations of Filament-Reinforced Composites for Aerospace Structural Applications," NASA CR-207, April 1965.

[9]Philips, D. C. and Tetelman, A. S., "The Fracture Toughness of Fiber Composites," *Composites*, Vol. 3, Sept. 1972, pp. 216–223.

[10]Sih, G., Paris, P. C., and Irwin, G. R., "On Cracks in Rectilinearly Anisotropic Bodies," *International Journal of Fracture Mechanics*, Vol. 1, 1965, p. 189.

[11]Harrison, N. L., "Splitting of Fiber-Reinforced Materials," *Fiber Science and Technology*, Vol. 6, 1973, p. 25.

[12]Griffith, W. I., Kanninen, M. F., and Rybicki, E. F., "A Fracture Mechanics Approach to the Analysis of Graphite/Epoxy Laminated Precracked Tension Panels," ASTM STP 696, 1979.

3. FIBER SYSTEMS

3.1 INTRODUCTION

General

The strength of a brittle material is controlled by the presence of flaws. As the probability of finding a flaw of a particular severity depends on the volume of material, a fiber with a low volume per unit length generally appears stronger on the average than the bulk material. This effect is illustrated in Fig. 3.1, where probability distributions of strengths are shown for a hypothetical brittle material tested in tension in bulk and in fibrous form. As indicated schematically, the bulk material has a lower coefficient of variation and a lower average strength.

For crystalline materials, more favorable orientation of the crystallites within the fiber may also give rise to increased strength and modulus compared to the bulk material. For example, polymer chains may be aligned along the fiber axis by a stretching process. Carbon fibers have graphitic crystals aligned along the fiber resulting from the structure of the polymer and retained during manufacture. The strength of the graphite crystal is anisotropic and this orientation of the crystallites gives high strength in the fiber direction.

The density of a material is of importance when considering its possible value as a reinforcing fiber. In many applications, the final weight is a dominant factor. This is especially true of aerospace applications and is gaining significance in some land-based uses. The density of a material is a function of the atomic number of the constituent atoms. Light fibers are therefore generally based on elements or groups of elements with low atomic number, including C, N, O, Be, B, and Si. Of these, even glass (basically Si and O) is considered heavy with a specific gravity of around 2.4–2.5.

The principal materials from which fibers can at present be produced economically for use in composites requiring high strength and stiffness are: silica-based glass, "electrical" (E-glass) and high-strength (S-glass) grades; carbon, graphitized grades; boron; aromatic polyamides or "aramids." (There is now an increasing interest in alumina and silicon carbide fibers, but these will not be discussed in the following.)

The potential of these materials is seen in Table 3.1 where the strength and modulus of fibers is compared with structural materials. In many applications, including aerospace uses, the weight of a component with adequate strength or rigidity is an important design factor. Materials for such applications can be compared on the basis of specific properties obtained by dividing the appropriate property value by the specific gravity.

When the fibers are embedded in a matrix that has little load-bearing capability, many composite specific properties are reduced in proportion to the volume fraction of fiber; in Fig. 3.2 a comparison is shown of the specific properties of some structural materials and unidirectional composites loaded in the fiber direction.

For further accounts of the matters discussed here, see, for example, Refs. 1 and 2.

Filamentary Single Crystals (Whiskers)

Some materials can be formed into filamentary single crystals with diameters in the range of 1–10 μm and length-to-diameter ratios up to 10,000. Under certain growth conditions, these materials have been found to have very high strengths. This has been associated with their crystalline perfection, which minimizes the occurrence of defects responsible for the low strength of the materials in bulk form. Whiskers with low density have been formed from B, B_4C, Al_2O_3, SiC, and Si_3N_4. Composites formed using such whiskers embedded in a suitable matrix have high strength and modulus. However, these materials are outside the scope of this chapter and will not be considered further; Ref. 3 gives a more detailed account.

3.2 MANUFACTURE OF COMMERCIAL FIBERS

Glass Fibers

Several types of glass may be used in the manufacture of glass fiber.[4] The main type is *E-glass*, originally intended for the electrical industry. E-glass

Table 3.1 Comparison of the Properties of Structural Materials and Fibers

Material	Specific Gravity	Tensile Strength, GPa	Elastic Modulus, GPa	Specific Strength, GPa	Specific Modulus, GPa
Aluminum L65 alloy (similar to 2.014-T6)	2.8	0.46	72	0.17	26
Titanium DTD 5173 (similar to Ti-6Al-4V)	4.5	0.93	110	0.21	24
Steel S97 alloy (similar to 4340)	7.8	0.99	207	0.13	27
E-glass	2.54	2.6	84	0.98	33
S-glass	2.49	4.6	72	1.85	29
Carbon (graphite), type I	2.0	1.9	400	0.95	200
Carbon (graphite), type II	1.7	2.6	200	1.52	118
Boron	2.5	3.5	420	1.40	168
Aramid (Kevlar)	1.44	2.8	130	1.94	90

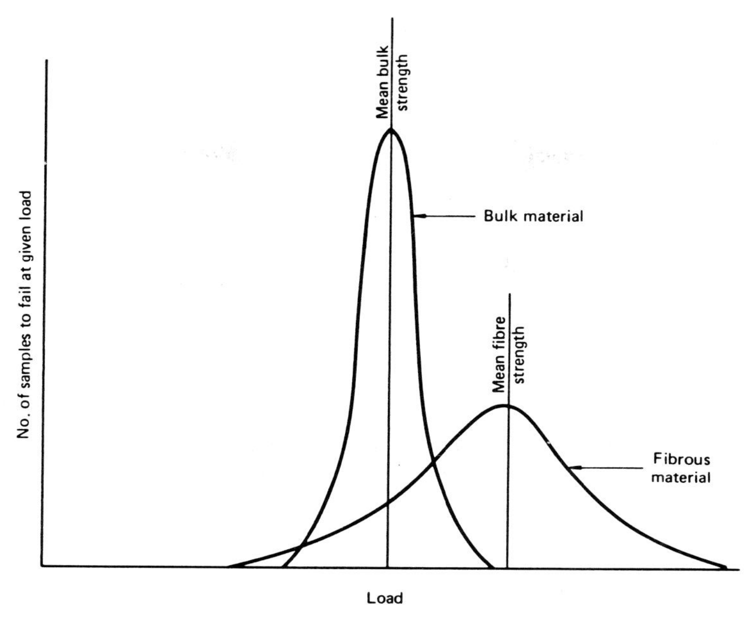

Fig. 3.1 Effect of sample cross section on distribution of strength.

is based on a mix of silica (SiO_2), alumina ($A1_20_3$), boric acid (H_3BO_3), and calcium carbonate ($CaCO_3$). Another type of glass is *S-glass* (S for strength). This is based on silica, alumina, and magnesia (MgO). The glass is usually premixed and supplied to the fiber manufacturer as glass marbles.

Fibers are formed by melting the marbles in a crucible at a carefully controlled temperature. The process is illustrated in Fig. 3.3. The melt, which has low viscosity, gravity feeds through a number of fine holes (200–400 holes of 1–2 mm diameter) in a platinum bushing. A winding device draws the filament bundle away at very high speed (50–100 m/s). The molten glass vitrifies within a few millimeters of the bushing and is cooled rapidly by an aqueous spray containing processing aids. These are discussed below. The fibers produced may be of any diameter, but most common filaments are about 12 μm.

Carbon and Graphite Fibers

Many forms of carbon fiber are available. Low-cost fibers are produced by pyrolysis of many organic, fibrous materials under controlled conditions. These fibers consist of largely amorphous carbon and have low strength and stiffness. They are used as a high-temperature insulation and for the manufacture of electrically conducting, general-purpose composites.

Advanced carbon (graphite) fibers can be produced from particular precursor fibers. Some manufacturers use polyacrylonitrile (PAN) or cellulose-based (rayon) fibers. Recent developments use fibers spun from pitch. The PAN process is illustrated in Fig. 3.4. The first stage involves a

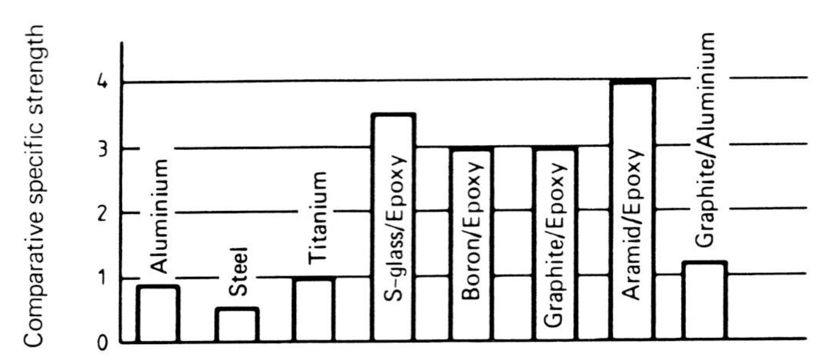

A comparison of composites and metals by specific strength (ultimate tensile strength/density)

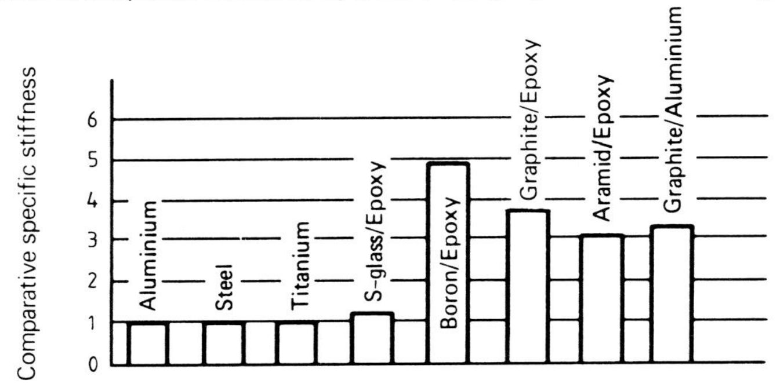

A comparison of composites and metals by specific stiffness (modulus/density)

Fig. 3.2 Comparison of specific strength and stiffness of composites and other materials.

controlled oxidation of a "tow" of about 10,000 filaments. Complex reactions between the aligned polymer chains of the precursor form cross-links and an extended carbon network retaining the original alignment along the fiber axis is formed. This stage is carried out with the fibers under tension. As this and the subsequent step require diffusion of waste gases to the fiber surface, the diameter of the fiber produced is limited. Fibers usually have a diameter of about 8 μm. The second stage is pyrolysis under an inert atmosphere. Most noncarbonaceous atoms are lost during this process. The final stage exposes the carbon filament to high temperature to facilitate formation of extended graphite ribbons, aligned along the fiber axis. The end result is a fiber with high modulus and strength. As the temperature of the final treatment is increased, the degree of "graphitization" increases and the fiber achieves a higher modulus. The strength of the fiber is affected by the number of flaws and defects present and their magnitude.[5] At higher heat-treatment temperatures, the flaws degrade the strength significantly. The fiber may be produced by controlling the process to optimize either modulus or strength, resulting in either high-strength (type HS or I) or high-modulus (type HM or II) fibers. Some economy can be obtained, for example, by reducing the final heat-treatment temperature, which results in production of a general-purpose fiber with reasonable, but lower, strength and modulus (type A).

A fourth variety of fiber is produced from a PAN precursor fiber having a "dog-bone" cross section. This overcomes, in part, the conflict during the optimization of the process and results in high-strength, high-modulus fibers. This may be related to a reduction in the distance necessary for evolved gases to diffuse out of the fiber without leading to high internal stresses.[6,7] The cross section, however, prevents the formation of high-fiber-content composites.

Boron Fibers

Boron fibers are produced by a vapor deposition process (see Fig. 3.5). A fine tungsten wire is heated by a passage of an electric current, in an atmosphere containing a boron trihalide (usually boron trichloride) and hydrogen. The reaction is

$$2BCl_3 + 3H_2 \rightarrow 2B + 6HCl$$

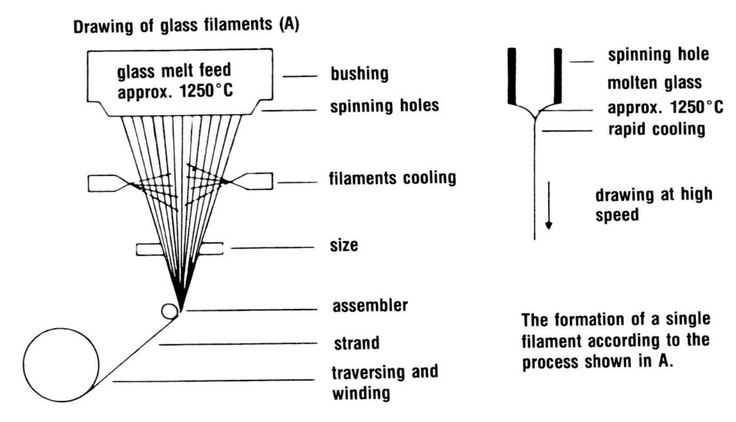

Fig. 3.3 Process of melt of spinning glass fibers.

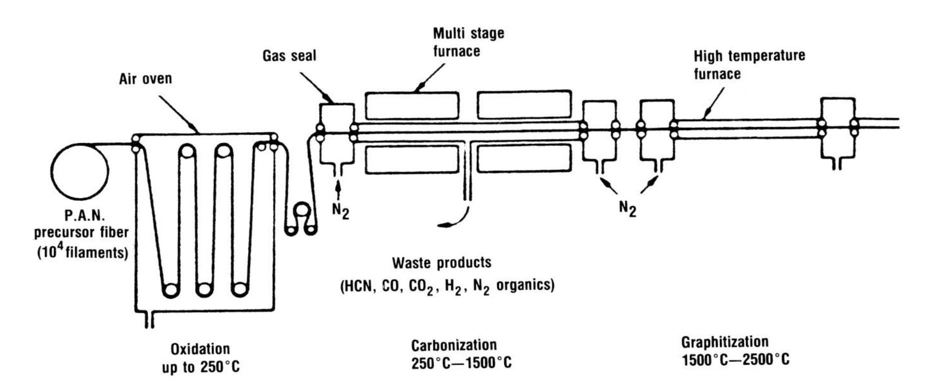

Fig. 3.4 Process used to manufacture carbon fibers.

With a carefully controlled temperature, the boron deposits evenly on the wire. The tungsten is largely converted to tungsten boride and remains in the fiber as a semiconducting core. The process is limited to single-fiber production. Pure tungsten is used as the core because of its high strength at elevated temperatures, but the initial diameter is limited by its strength. The boron filament produced must have a large diameter to ensure that the residual tungsten core has insignificant effects on the final fiber properties. Tungsten-cored fibers normally are produced with a diameter of about 125 μm. The bending stiffness of an individual fiber is dependent on the fourth power of the radius and thus thick fibers are difficult to manipulate into complex shapes. The tightest radius bend into which an individual fiber can be formed depends on the ultimate elongation of the fiber and its radius; boron composites will fracture if the radius of curvature is less than about 8 mm. It is thus not possible to weave boron filaments into a conventional cloth.

Recent developments use a carbon fiber as the core material. This gives a lighter fiber due to the low density of the carbon core; also, a smaller boron fiber diameter can be employed because there is now no need to overcome the problem of the density of the core.

Compressive properties of a unidirectional composite, in the fiber direction, may be limited by the columnar stability of individual fibers, by the fiber-matrix bond strength, and by the shear modulus of the matrix. Failure of carbon and glass composites in compression is initiated when the fibers buckle after the bond strength or matrix shear yield strength is exceeded[9]; boron fibers, because of their larger diameter, resist this buckling and, consequently, boron-reinforced composites have excellent compressive

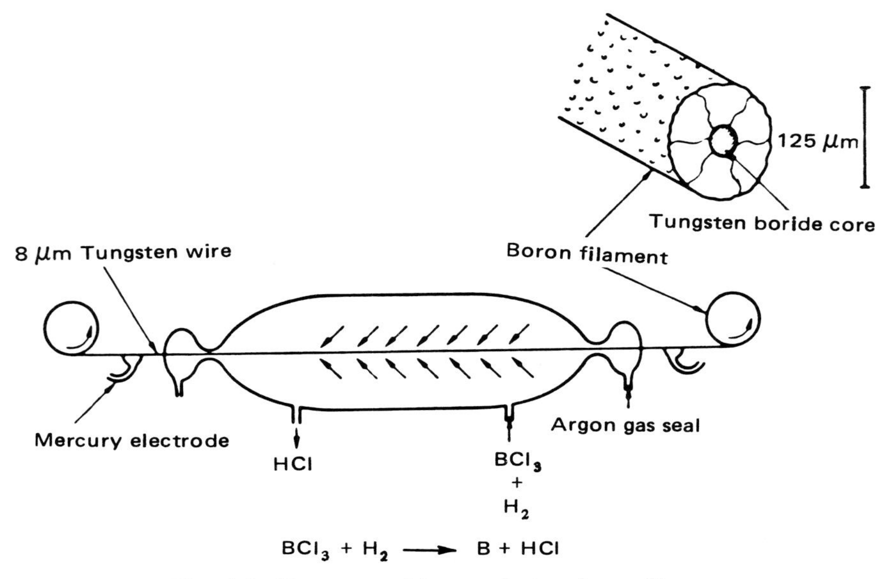

Fig. 3.5 Process used to manufacture boron fibers.

strength.[10] As boron is exceedingly hard (9 Moh), derived composites can be machined only with great difficulty using diamond-tipped tools.

Aramid Fibers

These fibers are the latest type to be released commercially. They are polyamide fibers closely related to conventional nylons, but in which the flexible aliphatic chains are replaced by aromatic rings. The polymer is formed by condensation of aromatic diacid derivatives with aromatic diamines. The only commercially available fiber is Kevlar marketed by Du Pont. This is based on the condensation product of terephthalic acid and paraphenylene diamine.[11]

These materials have long been known to have appropriate properties for use as reinforcements, but they are generally intractable polymers. The commercial availability of advanced solvents and spinning techniques made production of aramid fibers commercially feasible. The acid and amine are condensed in suitable solvents and formed into fibers by extruding a solution in sulfuric acid into a neutralizing bath when the fiber is precipitated. The polymer in the raw fiber is oriented along the fiber axis, as for most polymer fibers, by extended elongation of the fiber during and immediately following the spinning process. The fiber diameter is limited by the need to lose trace solvent from the finished fiber.

Three types of Kevlar are available, differing in the draw ratio of the final step. The first grade, called simply Kevlar, is intended for reinforcement of tires. The second, Kevlar 29, is used to reinforce rubber goods such as cordage and belting. The third, Kevlar 49, is used in engineering composites.

The structure of the fiber is itself a composite, with long stiff fibrils running along the fiber axis and embedded in a softer matrix. This microstructure imparts two important properties to Kevlar. The composite compressive strength is significantly lower than with other advanced fibers because the fiber buckling is enhanced by fibril buckling.[12] On the other hand, the fracture toughness of the composite is greatly increased, as considerable energy is absorbed in the failing composite by cracks running between fibrils. Failed specimens often give the appearance of broomsticks due to this failure mode. Machining of Kevlar composites requires careful design or modification of cutting tools.[13]

3.3 FORMS OF FIBER

Textile Terminology and Filament Forms

A number of special terms derived from the textile trade are used here. A single fiber is usually called a *filament*. A collection of filaments usually produced simultaneously is termed an *end* or sometimes a *strand*. Several ends may again be combined into a *roving* or a *tow*. Typical rovings may contain 50–60 ends and 3000–4000 filaments. If a strand is twisted to hold the filaments together it becomes a *yarn*, which may be further combined into *assembled*, *plied*, and *cabled* yarns. Complex yarns are not common in reinforcement applications, as excessive twist in the yarn reduces the ease with which resin can displace entrapped air (*wet-out*) from the filament bundle. However, some twist helps compact the yarn, resulting, ultimately, in a composite with a higher fiber content.

Woven Broadgoods

Both roving and yarn may be woven to give, respectively, *woven roving* and *cloth*.[4] Woven roving uses the fiber strength efficiently, but as the roving is not compact, generally derived composites have high resin contents. It is normally used to build thick sections rapidly, as it will wet-out easily with resin. Cloth is slightly weaker on the basis of fiber volume fraction than roving due, in part, to the increased damage caused during processing and, in part, to the effect of the fiber twist. Cloth allows a good wet-out by resin and, as it is compact, gives a high-fiber-content laminate. During the weaving process, woven rovings and cloth may be produced in a wide variety of forms. The ratio of the number of filaments across the cloth roll (*weft* direction) to the number along the roll (*warp* direction) can easily be varied from effectively unidirectional, with only sufficient warp filaments to hold the cloth together, to even warp and weft. Hybrid cloths are produced with different fibers in the warp and weft directions, e.g., graphite warp and glass weft.

The weaving pattern affects the way the cloth handles and the final composite properties. In plain weave, the weft alternately goes over and under successive warps. Special weave patterns, where the warp fibers cross different numbers of weft fibers result in *satin* and *crowsfoot* cloths.[4] In these, the cloth drapes over complex shapes more readily, but is easily disturbed to give alignment errors. The strength and stiffness tend to be higher because the fibers, on average, are straighter.

Nonwoven Broadgoods

Another form in which fibers may be supplied is *chopped strand mat* (CSM). This form is produced by chopping the fibers (usually glass) into lengths about 25–50 mm long and distributing them randomly on a flat surface. A water-soluble binder is sprayed onto the mat to hold the fibers together and the sheet is then compacted. In use, the binder is designed to dissolve in the resin system to give a random orientation of the reinforcement in the plane of the mat. Much chopped strand mat is designed for use with polyester and vinyl ester resins. The binder does not dissolve in epoxy resins and hence cannot be used with them. Some epoxy-compatible CSM is produced in Europe and the United States.

Some fibers may be produced in felt form, where short fibers are held in a light cloth with the fibers normal to the plane of the support cloth. This is occasionally used in an attempt to produce "through-the-thickness" reinforcement.

Some composites, particularly if based on thick woven roving, tend to have a resin-rich layer between the laminates. Felt or CSM may be used to reinforce this layer to prevent interlaminar shear failure.

Tapes

Automated machinery is often employed in composite production facilities, especially for aerospace applications. Many of these machines are designed to use tapes. The tapes may be unidirectional when the fibers or tows are aligned primarily along the tape direction or may be conventionally woven, usually using a plain weave. Unidirectional tapes may be held together with a light cross weave of the same fiber or a lower-cost fiber, may be supported on a light glass cloth (scrim), or may be on an inert backing sheet. Tapes are often supplied preimpregnated with the required resin system, which serves to hold the fibers together on the backing material.

Boron fibers, which are too stiff to weave, are normally supplied as prepreg tape on a glass scrim.

3.4 SIZES, FINISHES, AND FIBER SURFACE TREATMENTS

During processing of the original fiber into the form in which it will be finally protected by the matrix, considerable damage can be expected. This is particularly true of glass fibers, which are brittle and have a high coefficient of friction (for glass on glass, this is around unity). Mishandling results in the formation of microcracks in the surface that substantially weaken the brittle fiber. To reduce this damage, a coating is applied to the fiber as early as possible (on exit from the spinerette for glass). This coating material, termed a *finish* or *size*, serves several functions. It acts (1) as a binder to keep the filaments together (especially in CSM, see above), (2) as a lubricant to reduce the interfilament coefficient of friction, (3) as a coupling agent to promote durable bonding between the fiber and matrix, and (4) as a wetting agent to assist in resin wet-out.

Typical sizes for glass include amino silanes and proprietary chrome finishes (Volan). The size is often designed for use with particular classes of resin (e.g., amino silanes for epoxy resin). For many fibers systems, the size is applied early in manufacture and is designed to dissolve in the resin system during use. It may be suitable only for specific resins. Carbon fibers are often treated with a solution of unmodified epoxy resin or, more rarely, a stable epoxy resin curing agent system that acts as a compatible size. For woven fabrics, the mechanical degradation of the fiber can be severe and very-heavy-duty textile finishes are used that are usually removed after weaving to be replaced by a conventional finish as above.

For some fibers, particularly graphite, the surface may be etched in either the gas or liquid phase by oxidizing agents, including chlorine and bromine, nitric acid and chlorates. This improves the wettability by the resin and encourages formation of a strong, durable fiber matrix bond. Some strength improvement by removal of surface flaws has been demonstrated.[9,14]

References

[1] Langley, M. (ed.), *Carbon Fibers in Engineering*, McGraw-Hill Book Co., London, 1973.

[2] Galasso, F. S., *High Modulus Fibers and Composites*, Gordon and Breach, New York, 1969.

[3]Levitt, A. P., *Whisker Technology*, Wiley-Interscience, New York, 1970.

[4]"Woven Glass Fibre Fabrics for Plastic Reinforcement," British Standard BS 3396, Pt. 1, 1966. "Glass Fibre Rovings," British Standard BS 3639, 1969. "Woven Roving Fabrics of E Glass Fibre," British Standard BS 3749, 1974.

[5]Johnson, D. J. and Tyson, C. N., "The Fine Structure of Graphitised Fibres," *Journal of Physics*, Ser. D, Vol. 2, 1969, pp. 787–795.

[6]LeMaistre, C. W. and Diefenderf, R. J., "Origin of Structure in Carbon Fibers," *Carbon Symposium 1971*, edited by J.D. Buckley, American Ceramic Society, Chicago, 1972.

[7]LeMaistre, C. W., "The Related Effects of Microstructure and Thermal Expansion Anisotropy to the Strengths of Carbon Fibers," Weapons Research Establishment Tech. Memo. 587 (WR&D), 1972.

[8]Galasso, F. and Paton, A., "The Tungsten Borides in Boron Fibers," *Transactions of AIME*, Vol. 236, 1966, pp. 1751–1752.

[9]Weaver, C. W. and Williams J. G., "Deformation of a Carbon Epoxy Composite under Hydrostatic Pressure," *Journal of Materials Science*, Vol. 10, 1975, pp. 1323–1333.

[10]Greszczuk, L. B., "Micro-buckling of Unidirectional Composites," AFML TR-71-231, 1971.

[11]Dobb, M. G., Johnson, D. J., and Saville, B. P., "Compressional Behaviour of Kevlar Fibers," *Polymer*, Vol. 22, 1981, pp. 960–965.

[12]Kulkarni, S. V., Rice, J. S., and Rosen, B. W., "An Investigation of the Compressive Strength of Kevlar 49/Epoxy Composites," *Composites*, Vol. 6, 1975, pp. 217–225.

[13]"A Guide to Cutting and Machining Kevlar Aramid," Du Pont E. I. deNemours & Co., Wilmington, DE, Pub. No. 446, Dec. 1983.

[14]Johnson, J. W., "Factors Affecting the Tensile Strength of Carbon-Graphite Fibers," *Applied Polymer Symposia*, Wiley-Interscience, New York, No. 9, 1969, pp. 229–243.

4. RESIN SYSTEMS

4.1 INTRODUCTION

Functions of the Matrix

The matrix in a composite serves three main functions: it holds the fibers together, it distributes the load between the fibers, and it protects the fibers from the environment.

The ideal material from which a matrix is derived should be, initially, a low-viscosity liquid that can be converted readily to a tough durable solid, adequately bonded to the reinforcing fiber. While the function of the fibrous reinforcement is to carry the load in the composite, the mechanical properties of the matrix can significantly affect the way, and the efficiency with which, the fibers operate. For example, in a fiber bundle in the absence of a matrix, the straightest fibers bear a high proportion of the load. The matrix causes the stress to be distributed more evenly between all fibers by causing all of the fibers to suffer the same strain. The stress is transmitted by a shear process that requires good bonding between the fiber and matrix and also high shear strength and modulus for the matrix itself.

As the load is carried primarily by the fibers, the overall composite elongation is limited by the elongation to failure of the fibers. This is usually 1–1.5%. The significant property of the matrix is that it should not crack at this strain level. Resin systems for advanced composites tend to behave in a brittle manner with low strain to failure and with a high modulus compared to systems designed for nonreinforced applications.

Across the fiber direction, the mechanical properties of the matrix and the bond between the fiber and matrix dominate the physical properties of the composite. The matrix is much weaker and more compliant than the fiber and, hence, direct transverse loading of the matrix is avoided as far as possible during design of a composite component.

The matrix and matrix/fiber interaction can have a significant effect on crack propagation through the composite. If the matrix shear strength and modulus and the fiber/matrix bond strength are too high, a crack may propagate through the fiber and matrix without turning. The composite will then behave as a brittle material and failed specimens will show clean fracture surfaces. If the bond strength is too low, the fibers will act as a fiber bundle and the composite will be weak. At an intermediate bond strength, cracks propagating transversely through resin or fiber may turn at the interface and travel along the fiber direction. This results in absorption of considerable energy; composites that fail by this mode are tough materials.

Failed specimens will show considerable fiber pull-out and the fracture surface will be very rough, with lengths of bare fiber being visible.

A compromise is needed between a system with high bond strength (for efficient load transfer between fibers, but poor fracture toughness) and systems with lower bond strength (which will not be as efficient, but will have greater toughness). Composites used in situations where the stress levels and directions are not defined, or manufactured under conditions that reduce the fiber alignment accuracy, usually require softer, more forgiving matrices. Advanced composites used in situations where stress levels are well defined and manufactured with careful control of the fiber alignment can make better use of the ultimate fiber properties by using matrices with high modulus and high bond strength.

Some general information on resin matrices is given in Refs. 1–5.

Resin Types

The major types of resin used in production of composite components are epoxy, polyester and vinyl ester, and phenolic resins. The most common matrix is polyester resin, which is used in composite applications for fairly low-stress situations. The majority of advanced composite applications call for the use of epoxy resins. The search for improved matrices continues, especially to allow production of composites suitable for use at higher temperatures and with lower moisture sensitivity. The comparative physical properties of typical matrix resins are shown in Table 4.1.

4.2 EPOXY RESINS

General

Epoxy resins are a class of compounds containing two or more epoxide groups. The major types are formed by reacting polyphenols with epi-

Table 4.1 Comparative Properties of Resin Matrices

	Epoxy Resins				
Property[a]	RT Cured	Heat Cured	Advanced[b]	Polyester	Phenolic
Specific gravity	1.1–1.3	1.2–1.4	1.3	1.2	1.2–1.3
Tensile strength, MPa	50–70	70–90	60	50–60	50–60
Tensile modulus, GPa	2–3	2.5–3	3.5	2–3	5–11
Elongation to failure,%	2–6	2–5	2	2–3	1.2
Compression strength, MPa	80–100	120–130	300	120–140	70–200
Max. operating temperature,°C[c]	70–100	100–180	180	60–80	100–125

[a]Values quoted obtained using relevant ASTM standard methods.

[b]A typical grade recommended for use with advanced fibers.

[c]Quoted is the heat distortion temperature which measures the onset of the glass transition region, approximately.

chlorhydrin under basic conditions. If the polyphenol is diphenylol propane, the most usual epoxy resin is obtained. Trade names include such materials as Epikote or Epon 828 DOW DER 331, Araldite F, 6010, etc. The resin is supplied as a viscous, clear to light-yellow liquid.

Many other epoxy resins are manufactured for special purposes. The polyhydric phenol may be replaced by aminophenols or polyamine compounds. The triglycidyl ether of aminophenol and the tetraglycidyl ether of diaminodiphenyl methane (DDM) are finding widespread use in advanced composites and adhesives. Structures of resins and typical curing agents are shown in Fig. 4.1.

All epoxy resins can be self-polymerized using suitable catalysts, but the majority of applications make use of curing agents. The major classes of curing agents are aliphatic amines, which give cold curing systems, and aromatic amines and polyanhydrides, which give heat curing systems. The condensation is a strict chemical reaction and the rate cannot be adjusted by changing the mixing ratios (compare with polyesters, discussed below). The chemical reaction between the resin and the curing agent is strongly exothermic in the liquid state. As the condensation proceeds, gel particles precipitate from the reactant mixture. The reaction rate in the gel is controlled by diffusion of the reactants to active sites and the rate is reduced, compared to what it is in the liquid state. Eventually, all of the liquid gels, probably forming a dispersion of well-advanced particles in a less well-advanced matrix. Reaction continues in the solid state until the resin is sufficiently cross-linked to become a glass. Diffusion of the reacting species in the glassy state is very slow and the reaction effectively stops at that point.

The occurrence of rubbery and glassy states is characteristic of amorphous polymers. All polymers become glassy at low temperatures and, at high

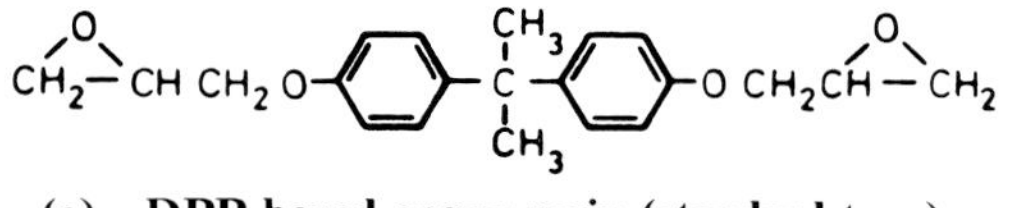

(a) DPP-based epoxy resin (standard type).

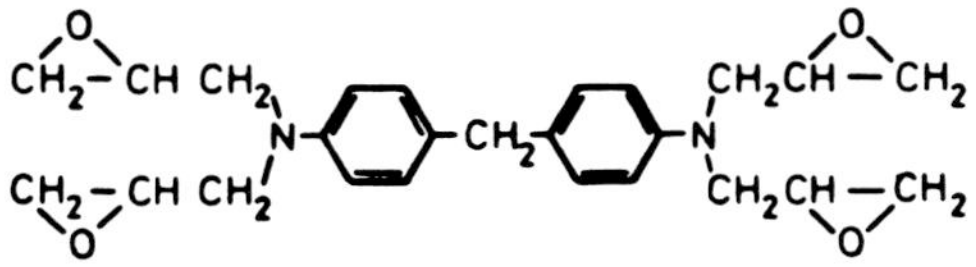

(b) Tetraglycidyl ether of DDM (advanced resin).

$H_2NCH_2CH_2NHCH_2CH_2NH_2$

(c) DET, room temperature curing agent.

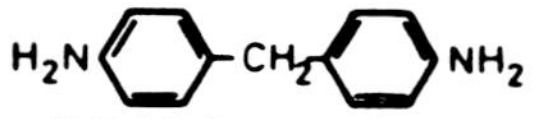

(d) DDM, heat curing agent.

Fig. 4.1 Structure of epoxy resins and curing agents.

temperatures, they may become rubbery. The temperature region over which the transition occurs is called the *glass transition temperature* T_g. The exact temperature is a function of how the measurement is made.

Most practical determinations of T_g involve stressing the sample and determining the effect of temperature. The speed of application of the stress and the rate of change of temperature have a pronounced effect on the observed temperature of the glass transition. Many rubbery plastics will fail in a glassy manner if tested at a high enough speed.

The glass transition temperature is a rough indication of the maximum operating temperature for that system. As the polymerization of epoxy resins stops in the glassy state, it is very difficult to design a system that will be capable of operation at much over the maximum temperature in the cure cycle. Systems cured at room temperature, using aliphatic polyamine curing agents, are not suitable for use at temperatures much higher than 50°C. With postcure at temperatures over 100°C, the maximum operating temperatures may be increased to 90–100°C. Systems cured with aromatic polyamines or anhydrides are usually cured at temperatures around 100–150°C, may often be postcured at 150–250°C, and have maximum operating temperatures in the range of 100–250°C.

Cure can be accelerated by the use of suitable catalysts, but the maximum rate is slower than for polyesters (see below). As the epoxide-curing agent is strongly exothermic (generates heat), the use of excessive quantities of catalyst or inappropriately high cure temperatures will result in thermal degradation of the matrix in thick sections.

Formulating with Epoxy Resins

The properties of the final cured matrix are partially defined by the choice of resin and curing agent. They may be further modified by a range of additives, including:

(1) Diluents added to reduce the viscosity before cure to aid in handling, wet-out, etc. (Usually, these cause decreases in the maximum operating temperatures.)

(2) Flexibilizers added to reduce the elastic modulus and increase the elongation to failure.

(3) Rubber-toughening agents that precipitate from the reacting matrix as rubbery particles, which modify crack propagation in the matrix.

(4) Inert fillers, including hollow fillers, added to alter density, cost, and effective matrix modulus.

Modification of properties is a specialized task and is not usually left to the user. However, commercial formulations are available from suppliers.

Environmental Durability

One of the functions of the matrix is to protect the fibers from the environment. Two components of the atmosphere that will degrade resins are ultraviolet (uv) light and moisture. All resins containing aromatic groups can absorb sufficient uv radiation to cause bond scission. Of the resins discussed, phenolics are most sensitive, followed by epoxy resins; polyesters are least sensitive. All resins may be protected from radiation by the use of surface coatings.

Attack by moisture is a more difficult problem. All fibers to which resins will form a bond are preferentially wet by water. Water, therefore, if allowed access to the fiber/resin interface, may result in debonding and a loss of composite stiffness and strength.

In the degradation by moisture ingress, the controlling factor is the diffusion constant of water vapor. As water is a very polar molecule, the diffusion mechanism involves hydrogen bonding with polar sites in the polymer molecule. Epoxy resins are the most polar of the normal resins as they contain hydroxyl groups, ether groups and C–N bonds. Polyesters are less polar due to the high concentration of polystyrene links and low concentration of ester links. Phenolics have the lowest polarity. Thus, water permeability is highest for epoxy resins and lowest for phenolics.

The equilibrium moisture uptake of the resin is also related to the diffusion constant and polarity of the resin. Water acts as a plasticizer of the resin system and slightly degrades the mechanical properties at room temperature. More significantly, it can reduce the glass transition temperature and limit the high-temperature performance of the matrix. As moisture is strongly bonded to the epoxy system even at high temperatures, it is lost slowly. High-performance aircraft may experience "thermal spikes" by cruising at high altitude (where skin temperatures may fall to around −50°C) and then sprinting to supersonic speeds (where aerodynamic heating may raise skin temperatures to well over 100°C in a few seconds). Special care needs to be taken that moisture absorption has not degraded the high-temperature performance of composites used in these situations.

It should also be mentioned that moisture trapped in voids and cracks can be vaporized under these conditions, which can result in matrix cracking. Similarly, low temperatures can freeze entrapped moisture and the resulting expansion can also initiate matrix cracking.

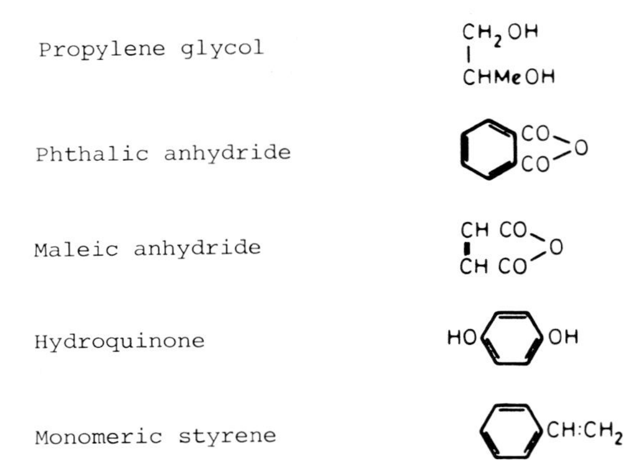

Fig. 4.2 Components of a polyester resin.

Advantages and Disadvantages of Epoxy Resins

The main advantages of epoxy resins are: (1) ability to formulate for optimum properties for particular applications; (2) control of fracture toughness; (3) moderate convenience to use; and (4) no volatiles, low shrinkage, good chemical resistance, good dimensional stability, good thermal stability, and high bond strength.

The main disadvantages are: (1) expensive compared to polyesters (specialty resins can be very expensive); (2) less convenient than polyesters due to relatively slow cure and high viscosity; and (3) limited resistance to some organic materials (particularly organic acids and phenols) and limited high-temperature performance.

4.3 POLYESTER RESINS

General

Polyester resins are formed by condensation of a mixture of dibasic acids with dihydric alcohols (glycols) or dihydric phenols. The dibasic acid mix contains unsaturated acids. The components are illustrated in Fig. 4.2. The condensation is carried out to give a prepolymer having a low melting point and good solubility in liquid styrene. The molecular weight of the prepolymer is a few thousand and it usually contains an average of five unsaturated links. The polyester is supplied as a solution in styrene monomer (35% styrene W/W). When a source of free radicals is added (the initiator), and often a catalyst (the accelerator) as well, the styrene polymerizes. The polymerizing styrene reacts with the unsaturated sites in the polyester to form a three-dimensional cross-linked network based in part on polystyrene and in part on the polyester prepolymer.

Types of Polyester

The major commercial variations of polyesters are based on modification of the polyester component by partial replacement of saturated acid or glycol by alternative materials. For example, resins of improved strength and durability are obtained by replacing the normal orthophthalic acid by isophthalic acid ("isophthalic polyesters") or by the use of diphenylol propane in place of some glycol ("DPP resins"). Another common variation is the use of adipic acid, which improves flexibility and increases the failure strain of the cured matrix.

Curing of Polyesters

The most commonly used initiator for cure of polyester resins is methylethyl ketone peroxide (MEKP), usually supplied as a solution in dimethyl phthalate. Cobalt naphthenate, supplied as a solution in naphtha, is used as an accelerator with MEKP. MEKP is an extremely hazardous material and can cause permanent eye damage, skin burns, etc., and can lead to serious fires and explosions if used incorrectly. Of particular importance is the admixture of the initiator and accelerator, which will spontaneously inflame and may explode. This occurs if the two materials are accidentally added successively to the resin without intermediate stirring.

The speed of the polyester polymerization may be controlled over a wide range by adjustment of the quantities of initiator (0.5–3% wt. resin) and accelerator (0.05–0.5% wt. resin) added.

The polyester polymerization is strongly exothermic and the use of high levels of initiator and accelerator will cause severe thermal damage in thick sections. The use of massive molds and the incorporation of fillers reduce the exotherm by increasing the system thermal mass.

Another common initiator is benzoyl peroxide (BzP), which is sold as a paste in dimethyl phthalate. The appropriate accelerator for BzP is a tertiary amine. It should be noted that the accelerators for MEKP and BzP are not interchangeable. Systems cured with BzP without an accelerator, or at very low accelerator contents, are stable at room temperature and are cured at elevated temperature. A range of initiators similar to BzP is available that will allow cross-linking to initiate rapidly at particular limiting temperatures; e.g., BzP is stable to 70°C, while ditertiary butyl peroxide is stable to 140°C. Heat-cured polyesters are generally used in matched die molding where fast cycle times are required.

Advantages and Disadvantages of Polyesters

The major advantages of polyesters are: (1) initial low viscosity that allows easy wet-out of reinforcement; (2) low cost (all raw materials are readily available and relatively inexpensive); (3) cure conditions that can be easily modified with little operator experience; (4) easy manufacture in a range of modifications for particular applications; and (5) good environmental durability.

The major disadvantages are: (1) high exotherm and high shrinkage on cure (both factors lead to a poor fiber/matrix bond strength due to in-built stress; (2) systems with adequate shear strength tending to be brittle and in which toughening additives appear ineffective; and (3) poor resistance to even very dilute alkali.

4.4 VINYL-ESTER RESINS

A resin system very closely related to polyesters is obtained by using unsaturated hydroxylic compounds (e.g., vinyl alcohol) in place of some of the glycol in a polyester and by eliminating the unsaturated acid. These materials are known as vinyl ester resins and are sold in styrene monomer solution and used in the same way as polyesters. The major advantage of these materials is an improved bond strength between the fiber and matrix. Generally, they are similar to polyesters.

4.5 PHENOLIC RESINS

General

When phenol is condensed with formaldehyde under alkaline conditions, polymerization occurs. If the system is carefully controlled, polymerization can be stopped while the polymer is still fusible and soluble. This prepolymer is termed a *resol*. It will polymerize under the influence of heat or of acidic or basic catalysts to give a densely cross-linked material of complex chemical structure. (See Fig. 4.3.) Water and other volatile by-products are formed, which requires that the polymerization should be carried out under high pressure to avoid the formation of a friable foam. Cured resol-type phenolics usually have a high void content.

If the prepolymerization is done under acid conditions, a different polymerization path is followed and a novolak resin is produced. This will not self-polymerize, but can be cross-linked under the influence of a complex amine, usually hexamethylene tetramine. (See Fig. 4.4.) Other materials, including paraldehyde, are occasionally used. Polymerization of the prepolymer is carried out under pressure as volatile by-products are also formed in this reaction.

Properties of Phenolic Resins

The phenolic prepolymers for use in composites are solids and are usually supplied in solution. Due to the dilution effect, the solutions are stable at room temperature. Fibers, usually in the form of cloth or mat, are impregnated with the solution and the solvent is evaporated. The resultant treated

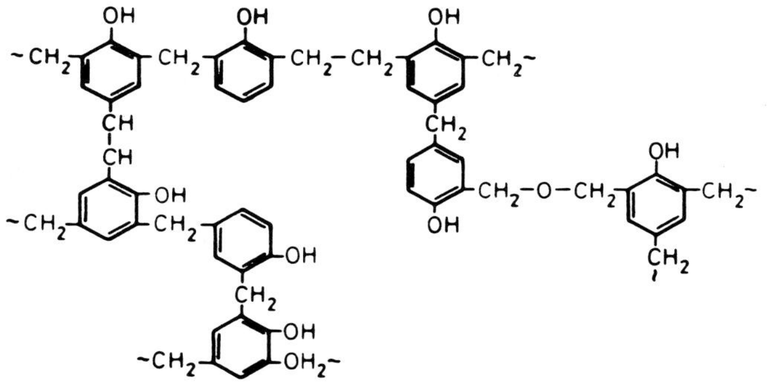

Fig. 4.3 Structure of a cross-linked phenolic resin (resol type).

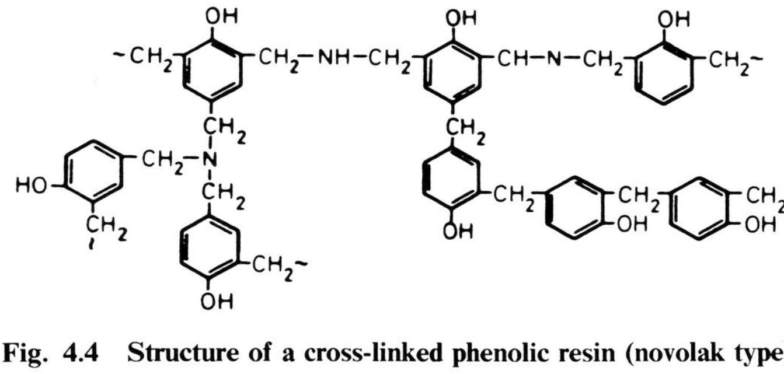

Fig. 4.4 Structure of a cross-linked phenolic resin (novolak type).

fabric may be heat treated to partially polymerize the resin system. This material is termed a *prepreg* (from "preimpregnated") and, if the resin has been heat treated, it is called a *B stage prepreg* (the *A stage* is a liquid or sticky solid prepolymer; *B stage* is a fusible solid with low tack; *C stage* is fully cured). Sheets of prepreg can then be assembled with the required orientation and consolidated under pressure during heat curing. While all heat-curing resins (including epoxy and polyester resins) are commonly used in this fashion, phenolic resins can be used only as prepregs.

Advantages and Disadvantages of Phenolic Resins

The principle advantage of phenolic resins is their excellent resistance to high temperature, especially under oxidizing conditions. This particularly affects their ablation properties, i.e., the speed at which they burn off when directly exposed to flame. Under these conditions, phenolics char readily and thus give a good yield of a superficial layer of porous carbon. This protects the underlying composite, while the carbon slowly burns away. Other resins usually have a poor char yield and burn away to gaseous products relatively quickly.

The disadvantages of phenolics include: (1) the difficulties in using them caused by the high pressures needed during polymerization; (2) their color (dark-brown to black); and (3) the fact that the mechanical properties of derived composites are lower than those for composites based on other resins due to the high content of the voids.

4.6 OTHER RESINS

A wide variety of resins has been developed in attempts to use composites at higher temperatures (e.g., 250–350°C). Commercially available in limited quantities are, for example, polyimides (PI), bismaleimides (BMI), and polybenzimidazoles (PBI). These are complex, highly aromatic materials frequently containing heterocyclic fused rings. They are produced from tetrafunctional aromatic acids and aromatic diamines (PI) or tetramines (PBI). Thermal treatment of this prepolymer, frequently in the presence of acid anhydrides and basic catalysts, results in the formation of cross-linked networks. The thermal resistance achieved to date allows continuous exposure to about 300°C. (They can withstand short exposures to up to 420°C.)

In most high-temperature resins, the curing reactions are intended to give additional rigidity to the network through interchain, molecular rearrangements at high temperatures. As the temperatures required for these complex reactions are so high, it is necessary to ensure high thermal stability in the partially formed prenetwork. Cure cycles tend to be complex in order to allow early polymerization and cross-linking to proceed to completion prior to exposure to the highest cure temperatures, when alternate cross-linking and molecular rearrangement reactions occur.

BMI resins can be handled in ways similar to those used with conventional epoxy resins. Such resins have improved thermal properties. PI and PBI resins are more difficult to use and require the prepreg process and high temperatures and high pressures for cure. PI has better temperature resistance than PBI, but is harder to use.

4.7 CARBON-CARBON COMPOSITES

A special type of advanced composite finding application in the aerospace field is the so-called *carbon-carbon composite*. This material is usually obtained by pyrolysis of a carbon-reinforced phenolic matrix under pressure. As described above, phenolic resins yield a high proportion of amorphous carbon char under such conditions, especially when air is excluded. The product is a carbon fiber reinforced amorphous carbon. This material tends to be very porous and must be reimpregnated and refined to reduce the void content. The finished material is capable of withstanding very high temperatures for prolonged periods if oxygen is excluded. It finds applications in brake linings, rocket exhaust nozzles, and other high-temperature applications.

4.8 FIBER-REINFORCED THERMOPLASTICS

Thermoplastics such as polycarbonates, polysulphones, and polyamides are available in grades reinforced with short glass or carbon fibers (1–25 mm). Usually, the fiber loading is low (less than 40% by volume) due to difficulties in processing the material by normal high-volume production techniques (such as injection or compression molding). In particular, the fibers tend to break up during processing and cause abrasive degradation in the process equipment. The mechanical properties of these materials are not generally adequate for structural applications in aircraft, particularly as most engineering thermoplastics have limited heat and solvent resistance.

Recent advances in thermoplastic technology have resulted in the development of high-temperature thermoplastics with excellent solvent resistance. Of particular importance is polyether-ether-ketone (PEEK).[6] Techniques have been developed for the manufacture of components from continuous fibers and such high-temperature thermoplastics. The usual method involves stacking alternate layers of dry fiber and thermoplastic film over a mold and compression molding the stack. PEEK is currently available as a preimpregnated thin film with continuous carbon or glass fiber reinforcement.

PEEK can be produced in amorphous and semicrystalline forms. It has a glass-transition temperature of 143°C and a melting point of 332°C. It may be processed at 300–400°C at pressures of 700–1500 kPa. Annealing at temperatures between the glass transition and the melting point results in development of the crystalline structure. A typical annealing cycle is 1 h at 200°C.

It has been found[6] that development of the semicrystalline structure is necessary for realization of the best chemical and thermal properties. Of particular interest are the very low moisture absorption of PEEK composites (0.4%) and the high fracture toughness compared to the normal brittle epoxy resin matrices. The low moisture uptake means that PEEK composites do not show a significant loss of mechanical properties under hot/wet conditions compared to hot/dry conditions. The high fracture toughness of the matrix means that PEEK composites do not appear to delaminate under low-energy impact conditions.

The overall thermal resistance is satisfactory up to a temperature approaching 100°C. This is adequate for many aircraft applications. Future developments will be directed toward upgrading the maximum operating temperatures.

4.9 QUALITY CONTROL OF RESINS, RESIN SYSTEMS, AND PREPREGS

When material is supplied in commercially pure form, as with conventional resins and curing agents, the incoming material can be tested using simple physical and chemical properties such as viscosity or epoxy equivalent weight. (For some special applications, a particular impurity may be significant, such as for electrical laminates when the presence of traces of chlorine can degrade the electrical properties of epoxy resins. This must then be tested for specifically.) The main problem with structural composites is that the incoming material is usually in prepreg form. The matrix is formulated using a range of epoxy resins and modifiers and is mixed with curing agent and accelerators. The mixed resin system is then used to impregnate the fiber system.

Until recently, the principal quality control techniques were to prepare finished composite specimens under controlled conditions and tests for their physical properties. Also required were tests on processibility, usually involving a determination of the flow properties of the resin. More recently, separative techniques have been improved using, for example, chromatography and it has become possible to specify the chemical constitution of the matrix and its degree of enhancement. It is still necessary to test for specific impurities, if known to be deleterious. These might include moisture content and residual solvent known to cause voids during normal resin processing.

References

[1] Saunders, K. J., *Organic Polymer Chemistry*, Chapman and Hall Ltd., London, 1973.

[2] Lee, H. and Neville, K., *Handbook of Epoxy Resins*, McGraw-Hill Book Co., New York, 1967.

[3] Boenig, H. V., *Unsaturated Polyesters*, Elsevier Publishing Co., Amsterdam, 1964.

[4] Lubin, G. (ed.), *Handbook of Fiberglass and Advanced Plastics Composites*, Van Nostrand Reinhold, New York, 1969.

[5] May, C. A., "Resins for Aerospace," ACS Symposium Series No. 132, American Chemical Society, Washington, D.C., 1980.

[6] Hartness, J. T., "Polyether-ether-ketone Matrix Composites," *SAMPE Quarterly*, Jan. 1983, pp. 33–37.

5. COMPOSITE SYSTEMS

5.1 INTRODUCTION

In the preceding chapters, the properties and nature of fibers and resins were discussed. This chapter will cover the characteristics of composite systems constituted from these materials. In Chap. 2, micromechanical analyses were used to predict composite properties from a knowledge of constituent properties and from an understanding of the fiber/matrix interactions in the "composite state." It is important to realize, however, that such analyses are approximate because they rely on very simplified models of very complex systems. Micromechanical analyses alone are not sufficient for design purposes and mechanical testing data from real composite materials are required. Therefore, emphasis is placed on testing procedures in this chapter. The elastic constants and strength values required for designing laminates are presented, the testing procedures for their determination are discussed, and the importance of the generally anisotropic nature of composites in the context of mechanical testing is demonstrated. Selected mechanical properties of several composite systems are discussed and compared with those of more conventional aircraft materials and, finally, the technique of hybridization of composites is introduced. References 1–3 provide a detailed coverage of the material presented in this chapter.

5.2 AVAILABLE FORMS OF MATERIAL

Advanced composites are usually supplied in the form of continuous tape or sheet. As shown in Fig. 5.1, the smaller-diameter fibers are grouped as tows or yarns, while larger-diameter fibers such as boron are in the form of monofilaments. With modern textile technology, a variety of woven forms are also available. Such forms allow easy combination of fiber types to produce hybrid composites such as aramid-glass, aramid-graphite, and graphite-glass.

In nonwoven unidirectional material, cross stitches of the constituent fiber or glass fiber are sometimes used to aid collimation of the fibers. Boron filaments are available only in unidirectional form (they cannot be woven because of their large diameter), but are easily collimated and are held in place with a light glass scrim. Although boron is most commonly supplied as a tape, "continuous" sheets are also available.

Fibers may be supplied in the dry condition or preimpregnated with resin, which aids in collimation and handling. The handling of these "prepregs" is also aided by partial curing (B staging) of the resin to provide tackiness. The major disadvantage of prepregs is their limited shelf life (even at $-20°C$) and the effect of resin cure advancement on processing variables.

Unidirectional and laminated composites are fabricated from stacked layers of prepreg by heat curing of the epoxy. Vacuum is generally applied during the initial stages of cure to aid in the removal of voids; then pressure is applied as the resin gels to remove excess resin. Usually a fiber volume fraction of 50–60% is most desirable. Further details of the fabrication process are discussed in Chap. 6.

5.3 PROPERTIES OF ADVANCED COMPOSITE SYSTEMS

General Comments

Experimental studies and theoretical analyses have shown that the properties of a fiber matrix system depend not only on the properties of the constituents, but also upon their manner of interaction in "combined form." It has been shown, for example, that broken fibers within a composite can still contribute to load bearing (Chap. 2) and that the microstructure of resins within composites (especially in close proximity to fibers) may be very different from that of the bulk material. Because of such complex features of a "composite system," the results of micromechanical analyses that rely on simplified models are only approximate and physical testing of the material is the only true guide for design.

Nonetheless, it is useful to use such analyses here, in a semiquantitative way, to explain some of the properties of the composite systems to be discussed. In particular, it should be noted that the exceptionally high

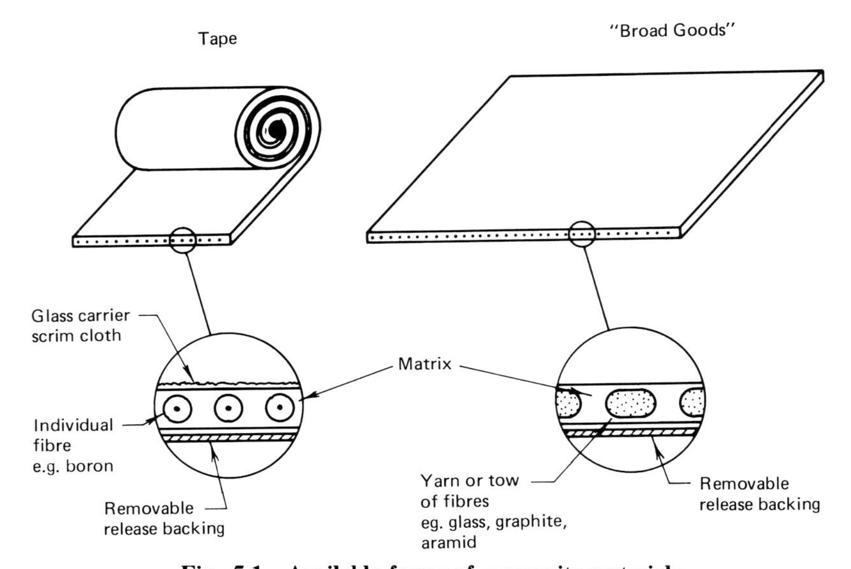

Fig. 5.1 Available forms of composite materials.

moduli and strengths of high-performance fibers are not fully realized in the composite system. This follows from predicted law of mixtures expressions of the forms (ignoring the contribution from the matrix),

$$E_1 \cong E_f V_f \quad (5.1)$$

and

$$\sigma_1 \cong \sigma_f V_f \quad (5.2)$$

where:

E_1 = modulus of composite in fiber direction
E_f = fiber modulus
σ_1 = tensile strength of composite in fiber direction
σ_f = fiber tensile strength
V_f = fiber volume fraction

Thus, in typical systems with about 60% fiber volume fractions, one could expect to achieve about 60% of the inherent fiber strengths and moduli in the longitudinal direction for a unidirectional composite. Values will be considerably less for multidirectional laminates and woven forms such as cloth.

The resin-sensitive, or "matrix-dominated" (transverse and shear), properties are obviously much lower than the "fiber-dominated" properties in advanced composites as indicated by relations of the form

$$1/E_2 = V_f/E_f + V_m/E_m \approx V_m/E_m \quad (5.3)$$

and

$$1/G_{12} = V_f/G_f + V_m/G_m \approx V_m/G_m \quad (5.4)$$

where:

E_m, G_m = matrix extensional and shear modulus, respectively
G_f = fiber shear modulus
E_2 = transverse modulus of composite
G_{12} = in-plane shear modulus of composite
V_m = matrix volume fraction

However, these properties are certainly of importance in design considerations.

In Table 5.1 are shown typical values for the systems of interest. It is evident that the fiber-dominated properties are far superior to the matrix-dominated properties. Longitudinal strength values are typically 10–100 times greater than corresponding transverse and shear values; longitudinal moduli are similarly much greater than transverse and shear moduli. It is interesting to note that Poisson's ratio (ν_{12}) may vary, depending on whether tensile or compressive stress is applied, and may also be a function of the magnitude of the applied stress; however, the latter effect is not significant in most structural applications in which the design limits of around 4000 microstrain are used to account for effects of environment, damage, etc. (Furthermore, Poisson's ratio may exceed 0.5 in anisotropic materials. For isotropic materials such as metals, Poisson's ratio is always less than 0.5.) Finally, it should be noted that, while it is reasonable to assume that moduli are equal under tension and compression (at least at low strains), it is clearly evident from Table 5.1 that strengths in tension and compression are markedly different.

Good design with composites utilizes as fully as possible the superior fiber-dominated properties. These properties are discussed in the following subsection, in which a comparison is made between advanced composites and more conventional aircraft materials.

Properties of Advanced Composites

In general, high-performance composites exhibit high strength and stiffness, low density, and good resistance to fatigue and corrosion, properties that make them very well suited to many aircraft and aerospace applications. Table 5.2 shows the markedly superior strength-to-weight and stiffness-to-weight properties of several composite systems compared to more conventional engineering materials. Boron/epoxy, for example, has four times the stiffness of an equivalent weight of steel and graphite/epoxy is five times stronger than the same weight of L65 aluminum alloy. It should

Table 5.1 Typical Properties of Unidirectional Composites

Property	Graphite/ Epoxy ($V_f = 60\%$)	Boron/ Epoxy ($V_f = 50\%$)	Glass/ Epoxy ($V_f = 45\%$)	Aramid/ Epoxy ($V_f = 60\%$)
Strength, GPa				
Longitudinal				
Tensile	1.1	1.3	1.1	1.4
Compressive	0.7	2.5	0.6	0.2
Transverse				
Tensile	0.02	0.06	0.03	0.01
Compressive	0.13	0.20	0.12	0.05
Modulus, GPa				
Longitudinal	130	200	40	80
Transverse	7	19	8	6
Shear	6	6	4	2
Shear strength, GPa	0.06	0.07	0.07	0.03
Poisson's ratio, ν_{12}	0.28	0.23	0.26	0.34

Table 5.2 Comparative Properties of Composites and Metallic Aircraft Materials[a]

Material	Specific Gravity	Tensile Strength, GPa	Tensile Modulus, GPa	Specific Tensile Strength, GPa	Specific Tensile Modulus, GPa
Boron/epoxy	2.0	1.49	224	0.73	110
Graphite/epoxy, type I	1.6	0.93	213	0.58	133
Graphite/epoxy, type II	1.5	1.62	148	1.01	92
Aramid/epoxy[b]	1.45	1.38	58	0.95	40
Glass/epoxy[c]	1.9	1.31	41	0.69	22
Steel[d]	7.8	0.99	207	0.13	27
Aluminum alloy[e]	2.8	0.46	72	0.17	26
Titanium[f]	4.5	0.93	110	0.21	24

[a]Unidirectional composites, 60% V_f. [b]Kevlar-49. [c]E-glass. [d]S97, similar to 4340. [e]L65, similar to 2014-T6. [f]DTD 5173, similar to Ti-6Al-4V.

be noted, however, that the properties shown here are for 0° unidirectional materials and are thus the maximum attainable values. In the multidirectional laminates encountered in practice, there will generally be substantial reductions from these values; for example, a typical laminate for a wing skin is likely to have an extensional modulus approximately one-half the unidirectional value.

It is also important to note that, while many design applications are based on strength-to-weight and stiffness-to-weight properties, these are by no means the only criteria. Material efficiencies or "indices of merit" can be calculated for various structural forms (see, e.g., Refs. 4 and 5); for example, the material efficiency of a thin panel which fails by buckling under compression is proportional to $E^{\frac{1}{3}}/\rho$, while that of a thin-walled tube which buckles under compression is proportional to $E^{\frac{1}{2}}/\rho$, where ρ denotes the density. The desirability of one material over another will thus depend on the details of the structure being considered. Even so, the strength and stiffness advantages of fiber composites on a weight basis are quite evident.

Typical tensile stress-strain curves are shown in Fig. 5.2. It can be clearly seen that glass/epoxy and aramid/epoxy, while having high tensile strengths, are not nearly as stiff as boron/epoxy and graphite/epoxy. A further disadvantage of aramid/epoxy is its very poor compression strength, but this is often overcome by hybridization with other fiber types, as will be discussed later. Certainly, the major advantage of aramid/epoxy and glass/epoxy is their low cost, aramid and glass fibers costing less than

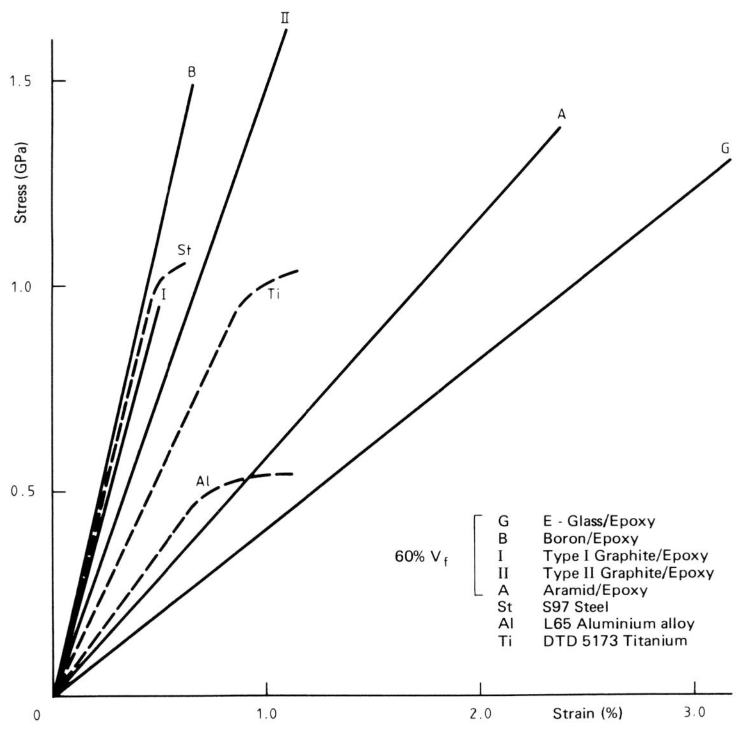

Fig. 5.2 Comparison of tensile properties of composites and metallic aircraft materials.

\$20/kg compared with graphite and boron fibers at \$100/kg and \$630/kg, respectively.*

Finally, it should be noted that advanced composites in general have much lower coefficients of thermal expansion than most metals. For example, for graphite/epoxy $\alpha = 0.5 \times 10^{-6}$ °C^{-1} compared with 23×10^{-6} °C^{-1} for aluminum alloys. This mismatch can cause problems with adhesively bonded composite/metal joints, as will be discussed in Chap. 8. The thermal expansion coefficient is also often markedly anisotropic in composites.

5.3 CHARACTERIZATION OF COMPOSITES

The mechanical properties of isotropic materials can be fully characterized by two elastic constants and a knowledge of tensile, compressive, and shear strengths. With composites, the situation is rather different; first, because of their anisotropy and, second, because they are generally used in

*The costs of graphite and boron fibers are presently undergoing changes. As graphite usage increases, its cost decreases, with some types of fiber now costing less than \$100/kg. As boron usage has not increased, its cost continues to rise.

the form of laminated structures. Macromechanical analyses can be used to derive the properties of laminated structures from those of the unidirectional ply (Chap. 7), so it is necessary only to characterize the unidirectional material. The elastic constants required for such characterization can be found from Hooke's law, which relates the stresses and strains in a material. For an anisotropic material in a general three-dimensional state of stress (Fig. 5.3a), Hooke's law can be written in matrix notation, in terms of the compliance matrix S_{ij} as

$$\begin{bmatrix} \varepsilon_1 \\ \varepsilon_2 \\ \varepsilon_3 \\ \gamma_{23} \\ \gamma_{31} \\ \gamma_{12} \end{bmatrix} = \begin{bmatrix} S_{11} & S_{12} & S_{13} & S_{14} & S_{15} & S_{16} \\ S_{21} & S_{22} & S_{23} & S_{24} & S_{25} & S_{26} \\ S_{31} & S_{32} & S_{33} & S_{34} & S_{35} & S_{36} \\ S_{41} & S_{42} & S_{43} & S_{44} & S_{45} & S_{46} \\ S_{51} & S_{52} & S_{53} & S_{54} & S_{55} & S_{56} \\ S_{61} & S_{62} & S_{63} & S_{64} & S_{65} & S_{66} \end{bmatrix} \begin{bmatrix} \sigma_1 \\ \sigma_2 \\ \sigma_3 \\ \tau_{23} \\ \tau_{31} \\ \tau_{12} \end{bmatrix} ; \; S_{ij} = S_{ji} \qquad (5.5)$$

Here there is a full coupling between the direct and shear strains ε and γ and the direct and shear stresses σ and τ.

Because of material symmetries, in the case of a unidirectional composite, with the stress/strain coordinate axes referred to the principal material

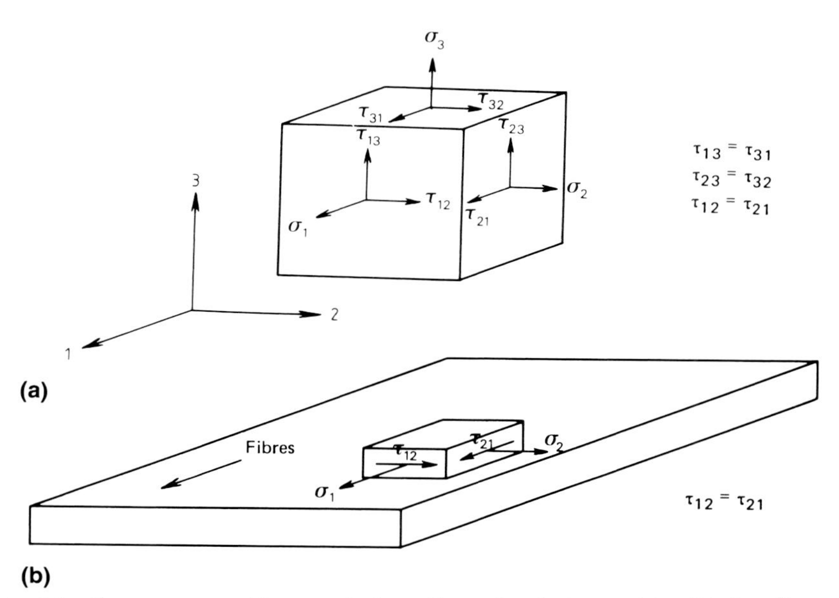

Fig. 5.3 Stress states: (a) general three-dimensional stress state; (b) two-dimensional plane stress state in a single ply.

directions, the shear strains γ are independent of the direct stresses σ and the direct strains ε are independent of the shear stresses τ, so that

$$\begin{bmatrix} \varepsilon_1 \\ \varepsilon_2 \\ \varepsilon_3 \\ \gamma_{23} \\ \gamma_{31} \\ \gamma_{12} \end{bmatrix} = \begin{bmatrix} S_{11} & S_{12} & S_{13} & 0 & 0 & 0 \\ S_{21} & S_{22} & S_{23} & 0 & 0 & 0 \\ S_{31} & S_{32} & S_{33} & 0 & 0 & 0 \\ 0 & 0 & 0 & S_{44} & 0 & 0 \\ 0 & 0 & 0 & 0 & S_{55} & 0 \\ 0 & 0 & 0 & 0 & 0 & S_{66} \end{bmatrix} \begin{bmatrix} \sigma_1 \\ \sigma_2 \\ \sigma_3 \\ \tau_{23} \\ \tau_{31} \\ \tau_{12} \end{bmatrix} \tag{5.6}$$

Any material having this type of stress-strain law is said to be "orthotropic."

A state of two-dimensional plane stress is usually assumed for individual layers within a laminate (Fig. 5.3b). Taking the plane of the laminate as the "1, 2" plane, and setting $\sigma_3 = \tau_{23} = \tau_{31} = 0$, the last equations reduce to

$$\begin{bmatrix} \varepsilon_1 \\ \varepsilon_2 \\ \gamma_{12} \end{bmatrix} = \begin{bmatrix} S_{11} & S_{12} & 0 \\ S_{21} & S_{22} & 0 \\ 0 & 0 & S_{66} \end{bmatrix} \begin{bmatrix} \sigma_1 \\ \sigma_2 \\ \tau_{12} \end{bmatrix} \tag{5.7}$$

(Note that τ_{31} and τ_{23} may well be present as interlaminar shear stresses; but, within the ply itself, they are negligible.)

The components of the compliance matrix can be expressed in terms of the more familiar engineering elastic constants, Young's modulus E,

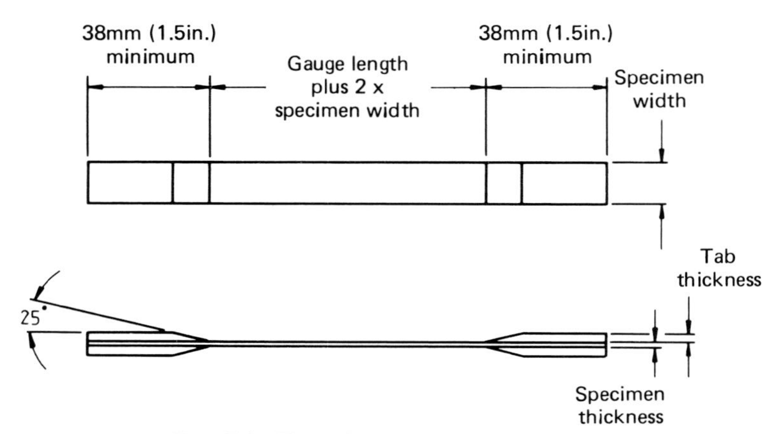

Fig. 5.4 Typical tensile test specimen.

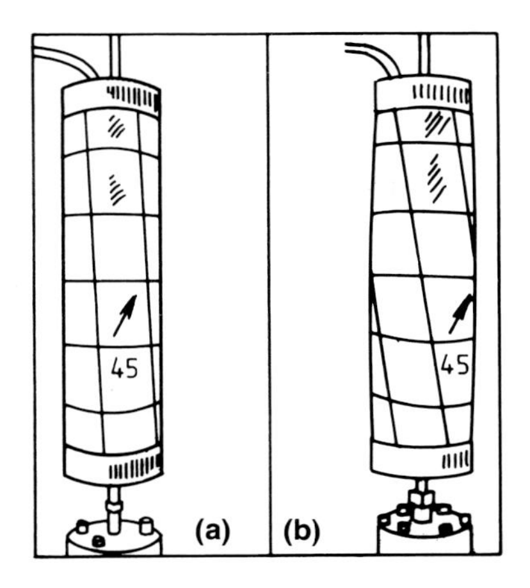

Fig. 5.5 Helically wound tubular specimen: (a) tension; (b) internal pressure.

Poisson's ratio ν, and shear modulus G, as

$$S_{11} = 1/E_1$$

$$S_{22} = 1/E_2$$

$$S_{12} = -\nu_{21}/E_2 = S_{21} = -\nu_{12}/E_1$$

$$S_{66} = 1/G_{12}$$

There are thus five elastic constants required to characterize the material, four of which are independent since $S_{12} = S_{21}$. The shear modulus G_{12} is also independent of the other elastic constants.

5.5 THE MECHANICAL TESTING OF COMPOSITES

Standard Testing Procedures

Fortunately, it is relatively straightforward to adopt many of the mechanical testing methods developed for isotropic materials in order to determine the mechanical properties of unidirectional composites in the principal material directions. Tensile moduli and strengths are usually determined using 0° and 90° parallel-sided specimens with bonded end tabs in standard test machines (Fig. 5.4). Bonded strain gages allow the determination of ν_{12} and ν_{21} at the same time. References 6 and 7 give details of standard testing procedures.

Compressive properties, particularly strength, are more difficult to determine because of brooming and buckling effects. Large, accurately bonded end restraints, short gage lengths, and possibly side restraints must be used on the test specimens.

Shear properties can be obtained (by the indirect use of laminate theory) using balanced ±45° tensile specimens[8] or by the torsion of a solid cylindrical rod with the fibers parallel to the longitudinal axis. This latter method can also be used to measure interlaminar shear response, at least to the limit of linearity.

The helically wound tubular specimen certainly represents the most unified characterization procedure (Fig. 5.5). All of the elastic constants and strengths can be determined[9,10] by applying an axial load with one end free to rotate, a torsional load with one end unconstrained in axial movement, and internal pressure. The only objection to using tubular or solid rod specimens is that their microstructures may not be representative of typical laminated panels. Off-axis properties can also be determined with tubular specimens.

Quality Control Tests

The most popular quality control tests are for flexural and interlaminar shear. They require only basic testing fixtures and use very simple specimen configurations with no end grips. Flexural tests can be done in three- or four-point bending and, for pure 0° and 90° specimens, can give valid estimates of strength and modulus.

The short-beam interlaminar shear test uses a three-roller jig and a specimen with a small span-to-depth ratio. Although it is convenient and useful for revealing defects such as interlaminar porosity and resin-rich layers, it cannot be used as a reliable measure of interlaminar shear strength because of the adverse effects of stress concentrations around the rollers and the nonuniform stress distributions. Testing procedures are detailed in Refs. 3 and 11.

Difficulties in Testing Composites

Although straightforward testing methods can be used for unidirectional composites when stress is applied in the principal material directions, difficulties arise with off-axis[†] testing and with multidirectional laminates. The major difficulty with off-axis testing is producing a uniform state of stress. Conventional gripping induces severe perturbations (bending and shear) into the stress field, an effect known as shear coupling (Fig. 5.6a). Severe errors would occur in calculated material properties using such a configuration and, for this reason, "fixed-grip" testing is suitable only for 0° and 90° or symmetric orthotropic laminates.[‡] The floating-grip" (Fig. 5.6b) does produce a uniform stress field, but measurements are obviously difficult and results must be carefully interpreted.[12,13] Similar coupling effects occur with unbalanced laminates (Fig. 5.7) and with off-angle tubular and cylindrical specimens if appropriate end constraints are not employed.

[†]An off-axis tension specimen is one in which the fiber direction does not coincide with the tension direction.

[‡]A symmetric laminate is one in which the midthickness plane is a plane of mirror symmetry; an orthotropic laminate is one in which for each ply with a fiber orientation of $+\theta$ there is a matching ply with a fiber orientation of $-\theta$.

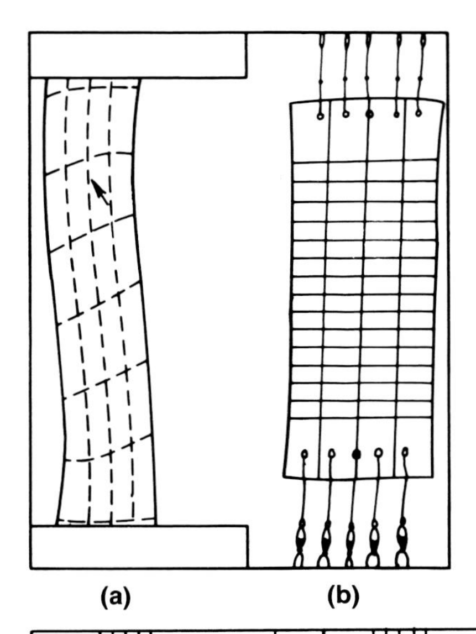

Fig. 5.6 Effects of end constraints on off-angle tensile response: (a) shear coupling effects; (b) uncoupled with "floating grip."

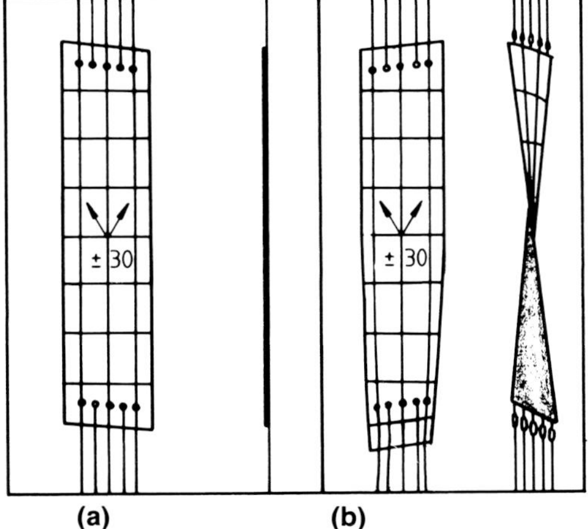

Fig. 5.7 Unsymmetrical laminate with "floating grip": (a) zero load; (b) tensile load.

Another problem, especially with bend tests, is the stress concentrations induced at the loading points. This, together with the tendency for nonlinear behavior, has restricted the use of such tests to quality control situations, although carefully interpreted results for 0°/90° laminates have given adequate results.[14]

Finally, it should be noted that, while it is easily realized that the sense in which direct stresses are applied (that is, either tensile or compressive) has a marked effect on material response, it is not as obvious that shear properties are affected in the same way. Consider, for example, the unidirectional 45° material under the two states of pure shear shown in Fig. 5.8. Clearly, the stress states felt by the fibers and matrix are quite different in the two cases and different failure modes will result.

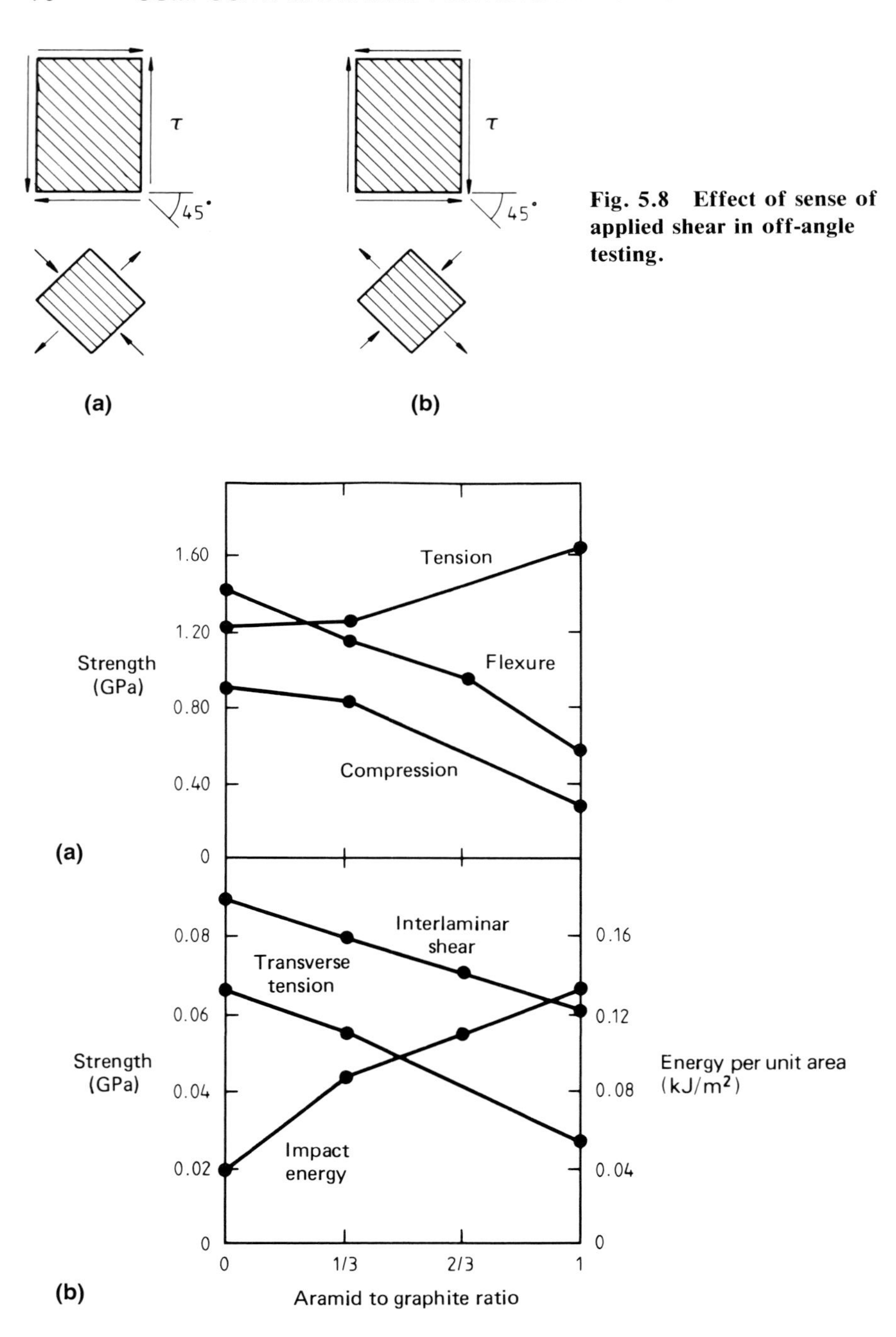

Fig. 5.8 Effect of sense of applied shear in off-angle testing.

Fig. 5.9 Variation of laminate mechanical properties with degree of hybridization for aramid-graphite fiber unidirectional composites.

5.6 HYBRID COMPOSITES

A major advantage of laminated structures is the ability to incorporate materials in various combinations, allowing the utilization of desirable properties of one material to overcome the deficiencies in another. A classical example of hybridization is the aramid-graphite hybrid, in which graphite provides the compression strength that aramids lack, while cost reduction is afforded by the aramids. The effect of hybridization on the mechanical properties is shown in Fig. 5.9a. Aramids are frequently used in hybrid structures where their high strain-to-failure capability provides improved impact resistance.[15] This may be seen in Fig. 5.9b where the improvement in Charpy impact energy with the increased use of aramid fibers is shown. Similar improvements in ballistic impact behavior have been reported with the use of hybrid systems.

References

[1]Ashton, J. E., Halpin, J. C., and Petit, P. H., *Primer on Composite Materials: Analysis*, Technomic Publishing Co., Westport, CT, 1969.

[2]Tsai, S. W. and Hahn, T. H., *Introduction to Composite Materials*, Technomic Publishing Co., Westport, CT, 1980.

[3]Whitney, J. M., Daniel, I. M., and Pipes, R. B., *Experimental Mechanics of Fiber Reinforced Composite Materials*, Prentice-Hall, Englewood Cliffs, NJ, 1982.

[4]Howard, H. B., "Merit Indices for Structural Materials," AGARD Rept. 105, 1957.

[5]Langley, M., *Carbon Fibres in Engineering*, McGraw-Hill Book Co., London, 1973.

[6]"Standard Test Method for Tensile Properties of Fiber-Resin Composites," ASTM D 3039-76, 1976.

[7]"Standard Test Method for Compressive Properties of Unidirectional or Crossply Fiber-Resin Composites," ASTM D 3410-75, 1975.

[8]"Standard Recommended Practice for Inplane Shear Stress-Shear Strain Response of Unidirectional Reinforced Plastics," ASTM D 3518-76, 1976.

[9]"Pagano, N. J., Halpin, J. C., and Whitney, J. M., "Tension Buckling of Anisotropic Cylinders," *Journal of Composite Materials*, Vol. 2, 1968, p. 154.

[10]Whitney, J. M., Pagano, N. J., and Pipes, R. B., "Design and Fabrication of Tubular Specimens for Composite Characterization," ASTM STP 497, 1972.

[11]Sturgeon, J. B., "Specimens and Test Methods for Carbon Fibre Reinforced Plastics," Royal Aircraft Establishment, Tech. Rept. 71026, Feb. 1971.

[12]Pagano, N. J. and Halpin, J. C., "Influence of End Constraint in the Testing of Anisotropic Bodies," *Journal of Composite Materials*, Vol. 2, 1968, p. 28.

[13]Wu, E. M. and Thomas, R. L., "Off-Axis Test of a Composite," *Journal of Composite Materials*, Vol. 2, 1968, pp. 523-526.

[14]Pagano, N. J., "Analysis of the Flexure Test of Bidirectional Composites," *Journal of Composite Materials*, Vol. 1, 1967, p. 336-342.

[15]Dorey, G., "The Use of Hybrids to Improve Composite Reliability," Paper presented at Technical Symposium on the Design and Use of Kevlar in Aircraft, Geneva, Oct. 1980.

6. COMPONENT FORM AND MANUFACTURE

6.1 INTRODUCTION

Fiber reinforcement is essentially a one-dimensional strengthening process. Since most components are stressed in more than one direction, a major function of the forming procedure is to orient the fibers in the matrix in the appropriate directions and proportions to obtain the desired mechanical properties. The forming process must also produce the shape of the component and develop the required properties of the matrix. In an ideal fibrous structure, the fibers may be aligned with the trajectories of principal stress and be concentrated in direct proportion to the local magnitude of the stress. This ideal is approached only by natural materials such as wood and bone.

The various manufacturing procedures for fiber-reinforced plastics may be classified according to the form of the reinforcement; Fig. 6.1 shows this classification and compares the resulting structures with other, more common, fibrous structures. At least in principle, all of the procedures based on continuous fibers (and some of those based on discontinuous fibers) allow close tailoring of the mechanical properties. These procedures, particularly those based on laminating, are used for manufacturing aircraft components from glass/epoxy, graphite/epoxy, aramid/epoxy, and boron/epoxy and will be the main subject of this chapter.

Typical generic aircraft components are listed in Table 6.1 with reference to the forming procedure employed. In general, laminated and filament-wound structures are relatively weak in the thickness direction and are thus limited to applications involving two-dimensional loading, such as those shown in Table 6.1.

Considerable efficiency is obtained by employing the reinforced material in the most highly stressed regions, for instance, in the upper and lower surfaces of those components subjected to bending. This is often achieved by using sandwich construction, with the composite forming the outer skins which are adhesively bonded to a honeycomb or plastic foam core. It is interesting to note that some natural fibrous materials (such as bone) may also employ this construction.

This chapter initially outlines general laminating procedures and then describes their application to advanced fiber composites. The process of filament winding and the related process of braiding are then described. Finally, the process of pultrusion is outlined. Joining, a very important aspect of manufacture, will be treated in Chap. 8.

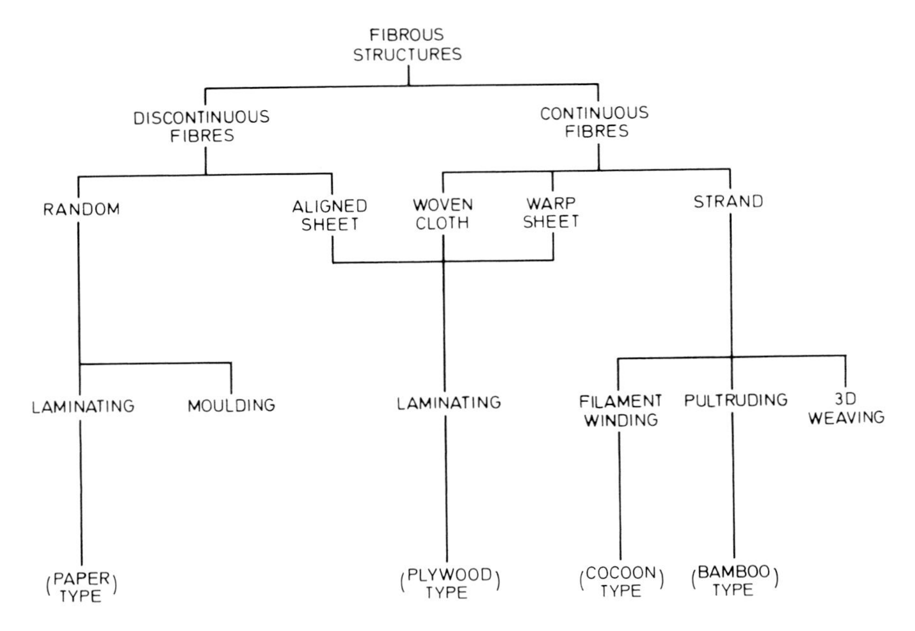

Fig. 6.1 Classification of manufacturing procedures.

6.2 AN OUTLINE OF GENERAL LAMINATING PROCEDURES

General

Most reinforced-plastic components based on continuous fibers are manufactured by some form of laminating procedure[1] in which sheets of fibers, either woven cloth or warp sheet (unidirectional fibers), are combined with resin and pressed against the surface of a mold. Heat, if required, is then applied to effect a cure. The various laminating procedures adopted are indicated in Fig. 6.2 and outlined below. The techniques described here are also applicable to laminates based on random mat materials and mixtures of random mat and cloth as employed, for instance, in the hull of a boat. Resin matrices are commonly polyesters or epoxies for low-cost applications.

Liquid Impregnation

In this procedure, layers of fibers are placed in a mold cavity that is subsequently evacuated; resin is then forced into the cavity by atmospheric pressure. Problems are caused by the tendency of the resin to foam under vacuum and also by the tendency of the resin to follow the path of least resistance through the fiber network. Considerable improvement in the process is obtained if the resin is applied under positive pressure, usually after the mold has been evacuated. Although this procedure is only occasionally used for advanced fiber composites, it is capable of producing high-volume fraction laminates with very good fiber distribution and low

Table 6.1 Typical Aircraft Fiber Composite Forms

Type of Structure	Typical Applications
Laminates	
Sheets, monolithic	Wing skins
Sheets, reinforced	Tail skins
Sandwich panels	Control surfaces
Shells	Fuselage sections
Beams	Spars
Complex forms	Airfoils
Filament Wound Fibers	
Closed shells	Pressure vessels
Open shells	Radomes
Tubes	Drive shafts
Secondary-formed tubes	Helicopter blades

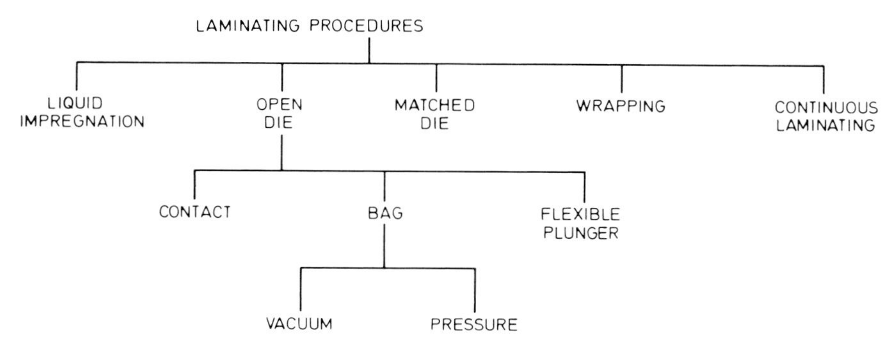

Fig. 6.2 Classification of laminating procedures.

void content. It is best if the fibers are precoated with a small amount of resin, since this aids in maintaining the alignment during impregnation.

Open-Die Molding

Open-die molding involves the use of only one mold surface, over which the layers of fiber are placed—or "laid up." If dry cloth is used, the resin may be applied by brushing or spraying. Alternatively, cloth or warp sheet preimpregnated with resin (prepreg) may be employed. Prepreg is used for most advanced applications; usually the resin is B-staged (partially cured or advanced) to provide the desirable tack and flow characteristics.

Various methods are employed to apply pressure to consolidate the lay-up; some of these are indicated in Fig. 6.2. In contact molding, which is generally employed only for fairly low-stress applications of glass/polyester composites, the pressure is developed by hand rolling over a sheet of plastic film placed over the surface of the lay-up.

The bag procedure involves the use of a flexible plastic membrane that is molded over the surface of the lay-up to form a vacuum-tight bag. In vacuum bagging, the bag is simply evacuated and atmospheric pressure used to consolidate the lay-up against the surface of the mold. Alternatively, gas pressure may be used to apply the consolidation pressure in an autoclave. In this case, the bag is evacuated initially to remove the air and volatile materials. Vacuum bagging is an inexpensive and versatile procedure; however, it can provide only limited consolidation pressure and may produce voided laminates due to the enlargement of the bubbles trapped in the resin in regions of low pressure. Autoclave procedures, described in detail later, are used to manufacture most of the high-quality laminates employed in the aircraft industry.

Pressure may be applied to the surface of an open mold by means of a flexible plunger. In this procedure, a solid rubber plunger is mounted in a press and used to force the lay-up against the wall of a female mold.

If required to effect the cure of the resin, temperature is applied to the open mold in various ways, including external methods such as hot-air blowers and ovens or internally by electric elements or steam pipes buried in the mold. Temperatures up to 180°C may be required for high-quality epoxy resin systems. The molds are usually of steel or aluminum alloy, particularly where internal heating is used. In order to reduce costs, open molds of plastic, wood, or glass/epoxy may be used, at least for short or preproduction runs; however, these tend to distort and are difficult to seal under vacuum conditions.

Matched-Die Molding

Matched-die molding involves the use of matching tool steel male and female dies that close to form a cavity of the shape of the component. The dies are internally heated, if required, by electric elements or steam pipes. The fiber layers are placed over the lower mold section and the two halves of the mold are brought together in a press. The thickness of the part is usually controlled by lands built into the mold. Advantages of matched-die molding are: very high dimensional control is possible, a good surface finish is produced on both surfaces, and production rates can be high. However, the cost of the matching dies (with hardened faces) is very high; also, the size of the available hydraulic presses used to apply the closing pressure limits the size of parts that can be produced.

Wet laminating procedures may be used, in which case the dry fiber is laid in the mold and the resin added subsequently. If random fiber mat is used, quite complex shapes can be produced; these are often preformed by spraying the fibers onto a male mandrel prior to placing the preform in the mold. Advanced fiber composite matched-die components are almost always based on prepreg starting materials.

Wrapping

Wrapping is an alternative procedure to filament winding for producing tubular components. Prepreg sheet, either warp sheet or cloth, is wrapped onto a removable metal mandrel and cured under pressure. Special machines are available to perform the wrapping operations. The pressure during an elevated temperature cure may be applied by shrink film (applied by a tape-winding machine), vacuum bag, or autoclave procedures. Alternatively, a silicon-rubber bladder may be initially placed over the mandrel prior to wrapping and then pressurized against an outer mold surface when the laminate has been wrapped in place.

Continuous Laminating

In this procedure, continuous flat-shaped or corrugated sheet is continuously fabricated. Sheets of cloth or mat, taken from a large roll, are passed through a resin bath and then brought together between plastic sheets by squeeze rollers that consolidate the layers and form the required section.

6.3 LAMINATING PROCEDURES FOR ADVANCED FIBER COMPOSITES

General

Advanced fiber composite aircraft components are usually made by a laminating procedure in which the plies of prepreg warp sheet or cloth are laid up, with the fibers at various prescribed angles, and consolidated against a mold surface under temperature and pressure.[2] The resin systems are generally epoxies; however, to achieve higher-temperature capabilities, resins such as bismaleimides may also be used. Fibers may be graphite, aramid, or boron.

The most common ply configuration employed consists of plies of 0°, ±45°, and 90°, where the angles are taken with respect to the principal load axis, e.g., the spanwise direction in the case of a wing. Essentially, the 0° fibers provide for the direct loads in the principal direction, the ±45° plies for the torsional loads about this axis, and the 90° plies for the transverse loads. The proportion of plies used in each direction depends on the particular application. Thus the skins for a sandwich panel used as a torsion box may consist largely of 0° and ±45° plies with a small number of 90° plies, whereas the skins for a sandwich panel used for a floor beam may have equal proportions of 0° and 90° plies. To avoid distortion, the plies are usually oriented symmetrically about the midplane of the laminate with equal numbers of +45° and −45° plies.

The laminating procedure is a very versatile method of manufacture. For example, it allows very wide variations in skin thickness; inclusion of local reinforcements, such as extra ±45° material around fastener hole regions; inclusion of metallic inserts for joining; forming of stiffeners; and formation of honeycomb core regions. An adhesive is often employed to bond metallic inserts into the laminate. The adhesive is cured simultaneously (cocured) with the resin in the laminate.

The laminate need not be made up of only one type of ply material. Hybrid constructions involving two or more materials may be employed to improve certain properties or simply to reduce costs. Thus, a graphite/

epoxy-aramid/epoxy hybrid may be used because it has a higher toughness than graphite/epoxy alone and a higher compression strength than aramid/epoxy alone.

The fabrication of advanced fiber composites, particularly those employed for flight critical components, calls for a very high degree of process control, very much higher than that normally associated with similar procedures for glass/polyester composites. Close control must be exercised over factors such as fiber content, orientation, and distribution; resin matrix integrity (freedom from voids, delaminations, cracks); resin properties (including degree of cure); and dimensional tolerances (including distortion).

In order to obtain the required level of control, prepreg warp sheet or cloth materials are used, employing open- or closed-die procedures. Open-die laminating, using autoclave procedures, is extensively employed, particularly for large components such as wing and tail skins, doors, fuselage sections, and spars. Matched-die molding is used only for long runs of small components or for fairly large complex components where very high dimensional accuracy is essential, e.g., fan or helicopter rotor blades. The thermal expansion of a rubber insert, or pressure expansion of a rubber bladder, may be used to produce internal pressurization of shell components, in combination with autoclave or matched-die procedures.

Prepreg Materials

Prepreg materials are based on a warp sheet or on woven cloth. Graphite and aramid warp sheet materials usually consist of axially aligned fibers held in position on release film by the tack of the impregnating resin. However, because of their large diameter (125 μm compared with 8 μm for graphite) and high modulus, boron fibers are held in place on a light woven glass cloth backing, which is incorporated into the laminate. Warp sheet and cloth prepreg materials are made by a specialist company to mold to a particular thickness when correctly processed; in the case of most graphite fiber warp sheet, this is about 0.13 mm. Then, the fiber volume fraction is about 58–64%.

There are available a wide variety of graphite and aramid woven fiber materials (boron is too stiff to weave into a cloth) having different weaves and tow sizes. Additionally, hybrid woven cloths are available in which two different fibers such as graphite and glass are combined in various arrangements. A typical example is a cloth having a graphite fiber warp and a glass fiber weft. Since cloth is relatively easy to handle dry, the preimpregnation may be carried out by the component fabricator.

In order to allow for the flow during the component manufacturing process, the resin content of a prepreg is usually up to 15% by weight more than required in the laminate. Until fairly recently, it was considered that resin flow is required to sweep out voids or bubbles. It is now considered by some manufacturers that, with good process control, excess resin is not required—particularly for woven cloth where the voids are not as serious since they do not lie in the interlaminar regions. Thus, the trend now is to use a low or zero-bleed prepreg.

Prepreg and cloth are available as tape about 75 mm wide or as sheet 1.2 m wide, contained between layers of release paper on rolls. Prepreg materials are usually stored at a temperature of about −20°C, at which their usable life is about 12–18 months.

Ply Cutting and Lay-up Procedures

Plies are cut from the prepreg material to the required size and, with the required fiber orientation, are then laid up onto the surface of the die or molding tool prior to consolidation under temperature and pressure. The cutting and lay-up operations are performed in a clean room, which is free of dust and held at constant temperature and (low) humidity. Various procedures are employed to cut the plies from wide sheet (obtained directly or formed from tape), including Gerber knife, laser, or water jet. These procedures, adapted from modern cloth cutting techniques in the textile industry, are generally numerically controlled. Each cut ply is held on a Mylar template that has index holes to allow correct positioning on the corresponding pegs on the mold and an identification number that allows correct sequencing of the ply in the thickness of the laminate. The ply is transferred to the laminate stack with a roller when the Mylar is positioned on the tool with the ply face downward.

Plies are formed from tape by a tape laying procedure, which may be hand controlled or partially or fully automatic. Tape is rolled onto the Mylar template in side-by-side strips to produce a sheet free of gaps and excessive overlaps. The automatic procedures often rely on an opaque marking tape attached to the Mylar that activates a cutter through a light-sensitive switch. A numerically controlled cutter (e.g., laser) is used to trim the cut edge to the correct contour. Alternatively, tape laying may be used to form the complete lay-up directly onto the mold surface without the need for separate operations. This procedure requires very uniform high-quality tape and very reliable machine operation to ensure that defective material, gaps, or overlaps are not incorporated into the lay-up.

Finally, for thick lay-ups (e.g., exceeding about 16 plies) several intermediate vacuum bagging operations may be employed to reduce the bulk (debulk) so as to minimize movement during subsequent molding. Vacuum bagging may also be used to preform laminates into deep contours.

Autoclave Procedures

An autoclave (Fig. 6.3) is essentially a large pressure vessel (pressurized with air, nitrogen, or carbon dioxide), which is internally heated and in which the lay-up is consolidated against an open die or mold. Typical temperatures are around 150°C and pressures about 1 MPa. A vacuum bag is formed over the lay-up and mold surface, primarily to allow pressure to force the lay-up against the mold surface. Prior to the application of gas pressure (from the time of lay-up), the bag is held under vacuum to keep the lay-up in position and to remove air and volatiles. When an intermediate temperature is reached, the vacuum is vented to the atmosphere and the temperature and pressure are raised to effect the cure. During the processing, it is most important to ensure that the resin is not allowed to gel under vacuum; such gelling causes a voided laminate due to the abnormal size of the gas bubbles under reduced pressure. However, too early an application of pressure may produce excessive resin flow, resulting in a resin-starved porous laminate.

The vacuum bag may be a thin impermeable plastic material, such as nylon (which is disposable) or a preformed silicone rubber (which is reusable). Butyl rubber paste is used to seal the disposable bag, whereas various types of molded-in ring seals are employed for the reusable bags. Several layers of materials are employed over the lay-up inside the vacuum bag (Fig 6.4), including a bleeder layer (glass or jute cloth) to distribute vacuum and absorb excess resin and a porous separator film (usually perforated plastic or PTFE-coated glass cloth) to prevent adhesion of the composite part to the bleeder layer. Any flaws in the vacuum bag system that may develop during the autoclave process are detected by leakage of the pressurizing gas through the vacuum line. The part is generally surrounded by a compressible (cork-neoprene) edge dam that serves to prevent sideways movement of the part, overthinning of the edges, and excess loss of resin from the edges. Thermocouples are sealed in position in various

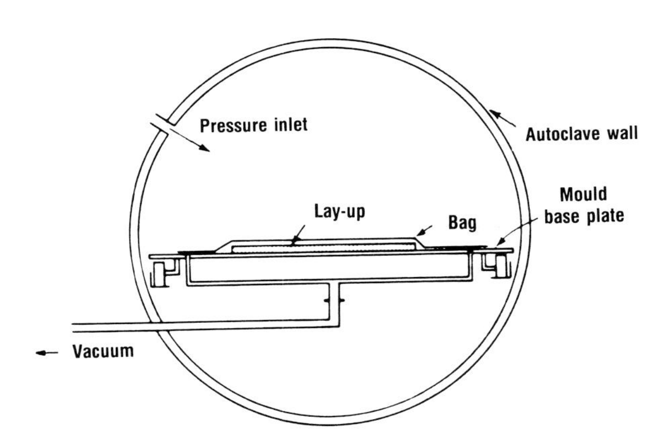

Fig. 6.3 Layout of an autoclave. (Taken from Ref. 2).

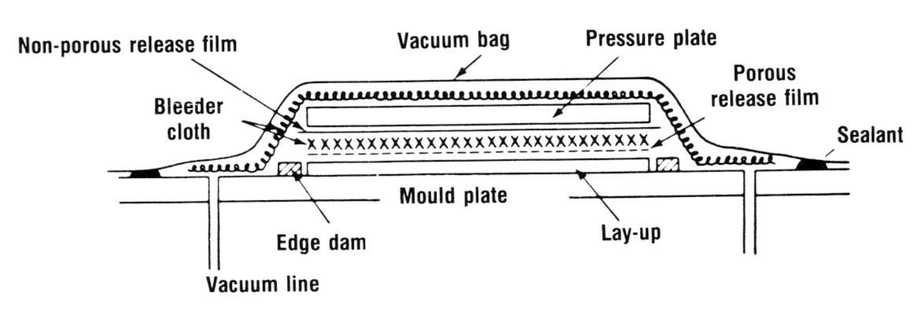

Fig. 6.4 Typical bag assembly for a flat panel molding. For other shapes, pressure plates are not normally employed. (Taken from Ref. 2.)

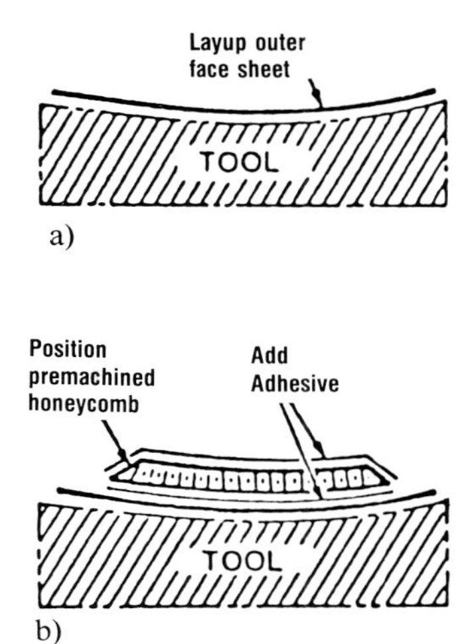

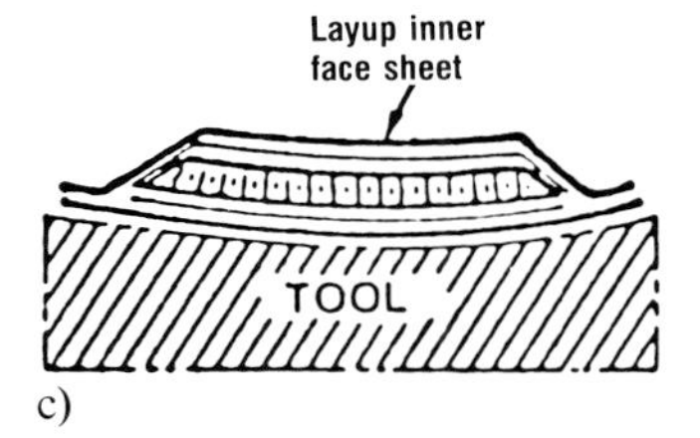

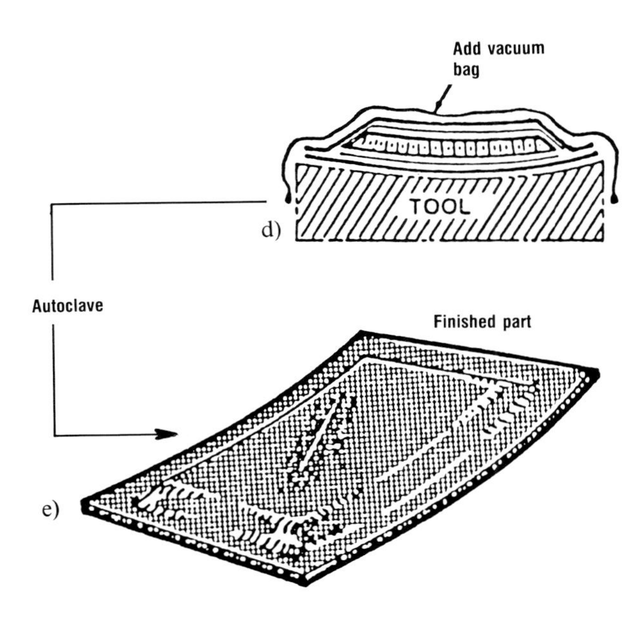

Fig. 6.5 Steps in the manufacture of a honeycomb panel by cocuring.

regions of the lay-up to determine the temperature distribution and to control the heating rate.

Autoclave tools must be vacuum tight, free from distortions under temperature and pressure, and have a low thermal mass to allow fairly rapid heating and cooling. The usual construction is a sheet steel fabrication; however, electroformed nickel is excellent for producing inexpensive, but durable, complex molds. Molds based on graphite/epoxy fiber honeycomb construction are also used. These have the advantage of thermal expansion compatibility with the graphite/epoxy laminates; they also resist distortion under pressure and, as a result of their low thermal mass, allow rapid heating and cooling.

Heat may be applied to the autoclave by using fans to circulate the pressurization gas over electrically or steam-heated elements. Alternatively, the molds may be internally heated from built-in electrical resistance elements. This latter method usually produces the best temperature distribution through the part.

Autoclave Manufacture of Some Typical Components

Manufacture of a honeycomb core component is illustrated in Fig. 6.5. In this case, the inner and outer skins are shown to be cocured to the honeycomb; however, in applications where optimum properties are required in the outer skin, the skin is precured.

Manufacture of sine-wave spars is a fairly complex operation, involving a combination of matched-die and autoclave molding; Fig. 6.6 shows the major steps.

Quite complex integral structures can be manufactured by autoclave procedures. For example, Rockwell International[3] has conducted fabrication studies involving the manufacture of tip-to-tip wing sections containing the lower skin and integral sine-wave spars. The procedure involved (1) preconsolidating and partially curing the spar (Fig. 6.7, left); (2) preconsolidating and partially curing the lower skin; (3) assembling the lower skin and spar (the spars are held in permanent location by barbed metal tacks); and (4) cocuring the spars and skins (Fig. 6.7, right). This procedure involves use of silicone rubber bagging and low-bleed (or prebled) prepreg material.

Matched-Die Molding, Cure Requirements

During matched-die molding, it is particularly important that pressure is applied at the correct gellation stage of the resin. The usual procedure is to apply a holding pressure for a dwell period at a single temperature (typically 150°C) until the onset of gellation and then increase pressure to close the dies. If pressure is applied too early, an excessive amount of the, still highly fluid, resin is expelled from the laminate, which then cures under low pressure once the dies contact on the lands. The result is a voided laminate with a very poor surface finish. If, however, the pressure is applied after gellation, it is not possible to close the dies and resin flow cannot occur to sweep out the voids; an oversize, resin-rich voided laminate results.

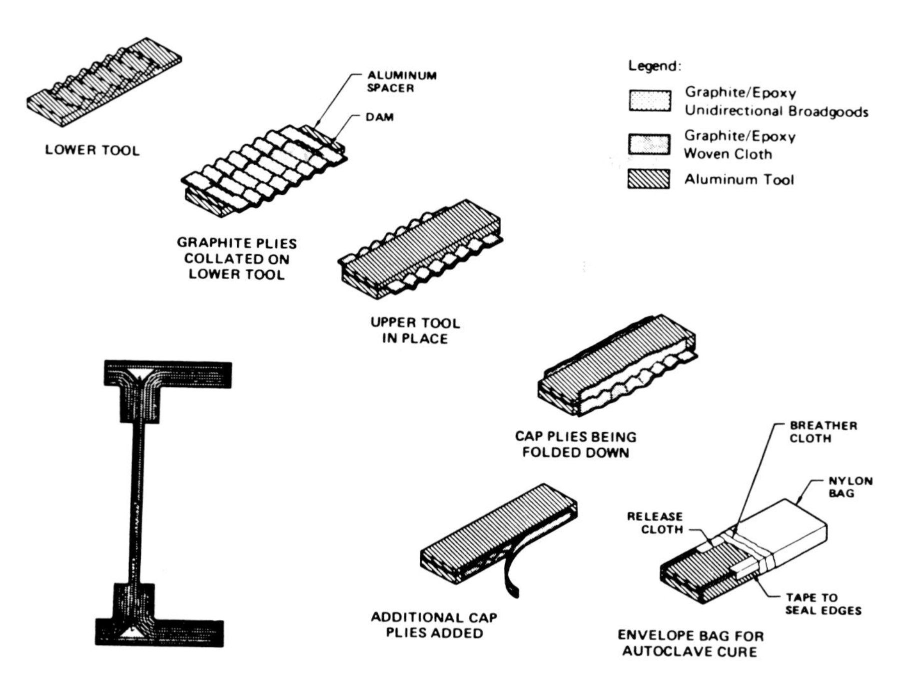

Fig. 6.6 Steps in the manufacture of a sine-wave spar.

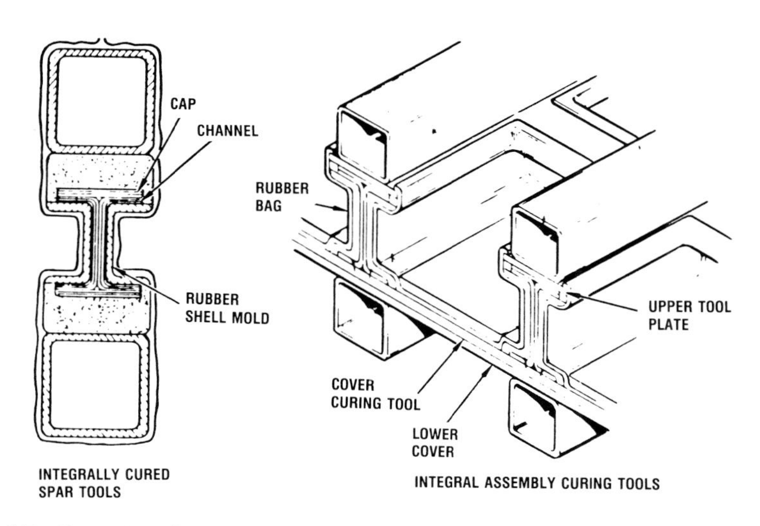

Fig. 6.7 Procedure for manufacturing an integral wing skin/spar structure. (Taken from Ref. 3.)

Expansion Molding Procedures

Thermal Expansion Rubber Techniques. Thermal expansion rubber molding utilizes the high coefficient of expansion of silicone rubber (approximately 10^{-4}°C^{-1}) and its thermal stability. This procedure can be employed to produce complex structures such as integrally molded skin boxes and complex forms such as sine-wave spars. Generally, elastomeric blocks are wrapped with the composite prepreg or the blocks are placed in position inside a formed lay-up. The assembly is then placed in a matched die, or an autoclave, which creates the expansion forces, allowing accurate formation of the outside face of the component. Often the elastomeric blocks are hollow to avoid development of excessive pressure during molding.

Zero-bleed prepreg is usually used, since it is difficult to allow for resin bleed in this type of molding. Vacuum bag debulking is employed to minimize the internal movement required during final molding.

Bladder Expansion Techniques. This procedure is similar to expansion rubber molding, except that the gas pressure is used to expand a rubber bladder wrapped with the prepreg against the cavity walls of a matched die. Alternatively, the bladder may be placed over a metal mandrel containing holes to allow access of the pressurization gas. This procedure is more controllable, but less versatile than expansion rubber molding. Bladder materials are usually cloth-reinforced silicone rubber or Teflon.

Expansion molding procedures are often used for the manufacture of complex components involving internal cavities and may be employed with either matched dies or in an autoclave. This procedure is also employed with filament winding, as described in Sec. 6.4.

Cure Monitoring

In order to ensure reproducible production of high-quality laminates (particularly in matched-die molding), a knowledge of the gellation time of the resin is required. Presently, this information is usually obtained by simple laboratory measurements that must be repeated as a given batch of material ages—a problem avoided by the use of fresh material. Various cure monitoring procedures have been developed to obtain information on the gellation point during the molding process.[4] These procedures are based on measurements of either resin viscosity, dielectric properties, or ion conductivity.

Measurement of the resin viscosity is the most direct method, but it is rather difficult to arrange, particularly with low-bleed resin systems, and is difficult to automate. Ion conductivity is based on the mobility of charged ions in the resin system; the mobility and, therefore, the conductivity of the resin falls as the resin cures.

The dielectric procedure is presently the most highly developed procedure. In this technique, two insulated metal electrodes are placed on either side of the laminate. The electrodes are connected to a dielectrometry instrument that measures the dielectric properties, particularly the loss tangent. The dielectric properties of the resin are a function of the ability of the dipoles in the resin system to rotate in phase with the electric field provided by the instrument. As the resin begins to gel, the dipole movement becomes more difficult. The change in the loss tangent is detected by the instrument and the indication is used as a signal to apply full pressure to the mold. Eventually, the presses and autoclaves could be controlled automatically by this procedure; however, it does not appear to be widely employed at present.

Quality Control

Prepreg. Prepreg materials are usually supplied by a specialist company to close specification. However, the aircraft manufacturer generally performs extensive tests[5] to establish that the following items conform to specification: fiber properties, fiber or resin content, resin/hardener formulation, residual volatile material, moisture content, flow properties, and degree of advancement (age) of the resin system.

Tests on the resin can be repeated through the refrigerated storage ($-20°C$) period to ensure that the prepreg is used before it overages.

Processing. During manufacture of the component the temperature in various regions of the laminate is recorded and plotted, together with details of the pressure cycle such as the leak rate of the pressurization gas through the vacuum bag, which is a measure of the integrity of the bagging system.

Table 6.2 Manufacturing Faults in Fiber Composite Laminates

Plain Laminates
Voids
Delaminations
Disbonds (e.g., from metallic inserts)
Foreign body inclusion (e.g., release film)
Resin starved areas
Resin rich areas
Incomplete resin cure
Incorrect fiber orientation (including wavy fibers)
Incorrect ply sequence
Fiber gaps
Wrinkled layers
Poor surface condition (e.g., poor release from mold)
Tolerance errors
Sandwich Panels (in addition to above)
Poor core splice
Disbonds from skins
Crushed core
Core gaps

Test coupons are laid up and processed with each major component for evaluation, usually of interlaminar shear strength and flexural strength. In critical applications, structural details may also be made and evaluated under representative stress and environmental conditions.

Final Laminate. Some of the types of manufacturing faults that can occur in the laminate component are listed in Table 6.2 Superficial faults, dimension errors, and distortion are readily found by inspection. Extensive nondestructive inspection (NDI) measurements are performed to inspect for internal flaws. The main techniques employed are C-scan ultrasonics in a water bath, or employing a water jet, to detect voids and delaminations and radiography to detect foreign objects (such as release film) and to examine the condition of honeycomb core regions.

6.4 FILAMENT WINDING

General

Filament winding[6,7] is the second major process for producing high-performance fiber composite structures. It is mainly employed to produce pressure vessels and other shell-like structures. The process involves winding continuous fibers, combined with resin, over a rotating mandrel, which is subsequently removed after cure of the composite.

The major advantage of filament winding is that the fibers can be laid under controlled tension precisely onto the curved mandrel surface. Since, with some winding patterns, the fibers can be made to follow a geodesic path (the shortest distance between two points on a curved surface), it is possible to arrange for very efficient utilization of the fibers. This is particularly useful in a pressure vessel where, ideally, all of the fibers can be uniformly loaded in tension.

The filament winding procedure is a much less versatile fabrication process than the general laminating procedures described earlier, particularly for complex forms with varying thickness and fiber orientation. However, variants of the procedure have been developed that allow formation of a variety of shapes which are not simple solids of revolution and could ultimately extend the capability of the procedure to the production of wing torque boxes and fuselage sections.

Winding Patterns

The basic winding patterns employed are illustrated in Fig. 6.8. The longitudinal, or polar, winding pattern is obtained by winding over the poles of the mandrel. This method is particularly suited to winding enclosed vessels; however, to obtain the required hoop strength in a pressure vessel, the polar wind must be overwound with a number of hoop layers, in the same ratio as the biaxial stress ratio. The helical winding pattern is formed by a winding machine with a lathe type of action (Fig. 6.9) using a controlled out-of-phase movement between the mandrel and the winding head to obtain the correct fiber distribution. Helical winding may be used

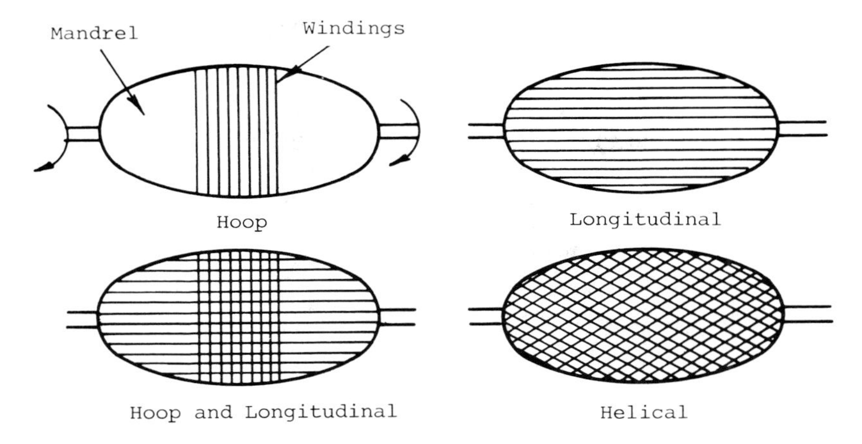

Fig. 6.8 Filament winding patterns.

alone, in which case the optimum winding angle for a pressure vessel is 54.75°, or with hoop windings. The helical winding pattern produces crossovers; this is a disadvantage because the local bending at these points increases fiber stresses. However, the interweaving improves interlaminar strength. Helical winding produces an ideal fiber geometry, particularly over end closures, since the fibers are wound on geodesic paths.

Design of filament wound structures is often based on netting analysis, which assumes that the fibers are uniformly stressed in tension and provide all of the longitudinal stiffness while the matrix provides only the shear stiffness. However, a more realistic procedure is to use laminate theory, which more correctly accounts for matrix contributions.

Manufacturing Procedures

The winding machine (Fig. 6.9) consists of a mandrel and a head to feed the reinforcement. Ancillary to this are devices to control the tension in the filament tows that are taken from several spools or creels.

In the wet winding procedure, dry fibers are passed through a heated resin bath. Rollers are used to remove the excess resin, force the resin into the fiber strands, and form a flat tape. In order to improve handleability and prevent damage, the fibers are usually coated with a light size compatible with the matrix resin; often the size is a part-cured resin of a composition similar to the matrix resin. Alternatively, a prepreg may be used, in which case the mandrel must be heated to obtain tack and resin flow. Use of prepreg allows high winding angles because of the high tack that can be developed and avoids resin handling problems.

Mandrels may be constructed of one-piece steel or aluminum if withdrawal is possible. Where simple withdrawal is not possible, various types of removable mandrels are employed, including those made of low-melting-point metal or soluble plastic. Inflatable and collapsible mandrels are also used.

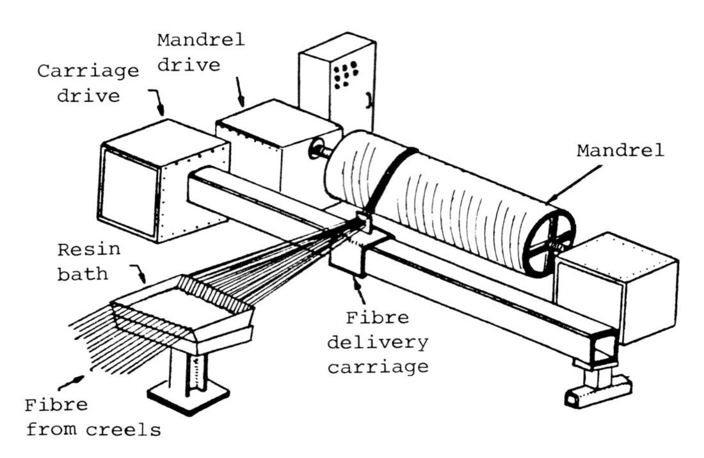

Fig. 6.9 Filament winding machine.

Filament-wound vessels can leak under high pressures through voids and cracking in the resin matrix. This problem is generally overcome through the use of metal or rubber internal liners that are incorporated by placing them on the mandrel prior to winding.

Metallic end closures are usually required in pressure vessels. These are placed on the mandrel and incorporated into the structure during winding.

Glass, aramid, and graphite fibers can all be successfully filament wound. However, graphite fibers can cause severe handling difficulties because of their tendency to break up during passage to the mandrel. This problem results from their high stiffness and high coefficient of friction. Frictional problems can be greatly reduced by the use of a thin resin coating or size on the fibers.

Secondary Molding

Filament winding can also be used to produce components that are not solids of revolution. Components for helicopters and remotely piloted vehicles have been made by this process, which initially involves winding onto an inflatable plastic or rubber mandrel. After winding, the mandrel (covered by the still wet winding) is deflated and placed in a matched-die mold; the mandrel is then reinflated to pressurize the composite and heat is applied to effect a cure. Even structures with a honeycomb core can be made by this procedure by employing two stages of winding.

Manufacture of Helicopter Blades

Filament winding procedures, combined with secondary molding, are now generally employed to produce the spars for helicopter blades. The winding and molding procedures involved in a particular case,[8] shown schematically in Fig. 6.10, involve polar winding the 0° fiber spar caps,

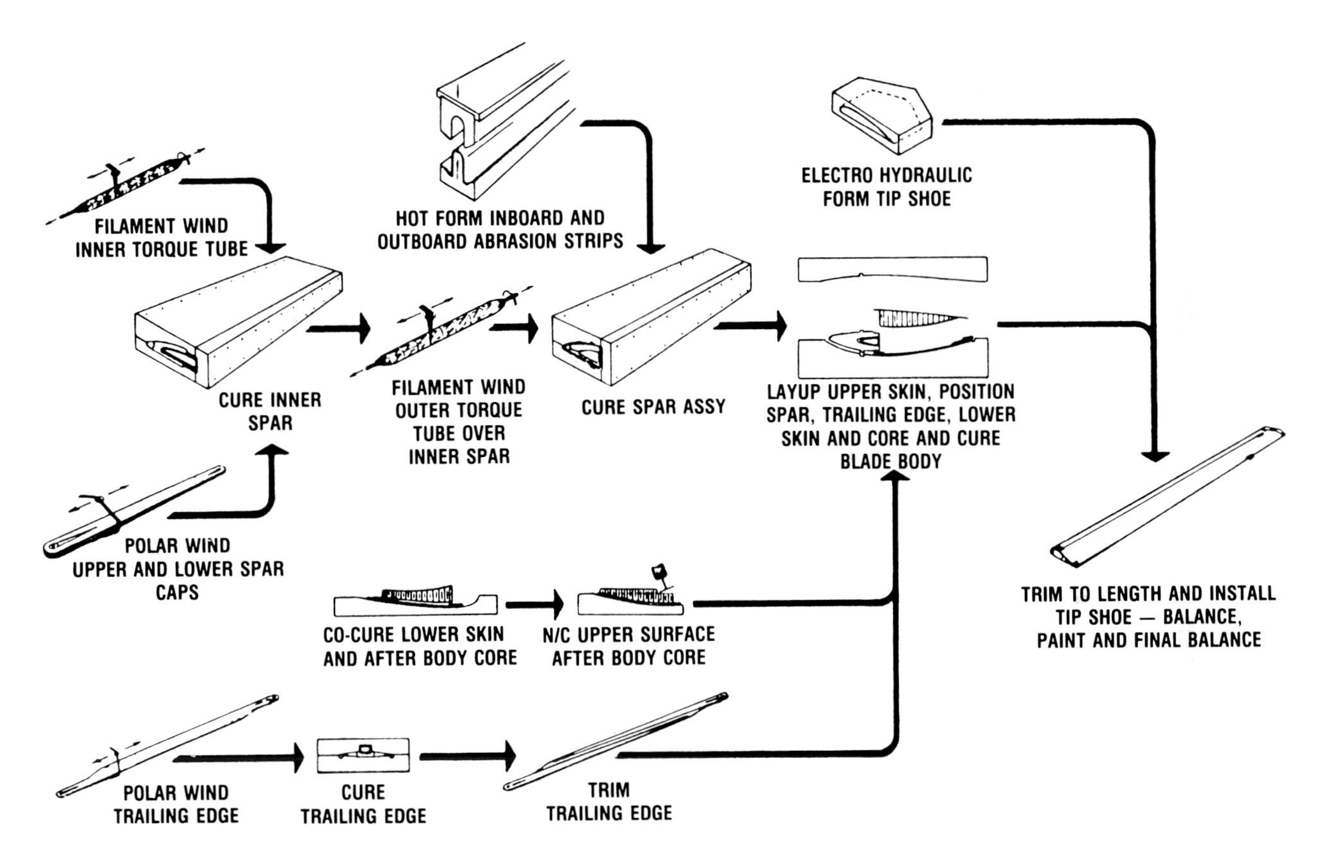

Fig. 6.10 Procedure for manufacturing a helicopter blade. (Taken from Ref. 8.)

including wrap around roots; helical winding ±45° inner and outer torque tubes for the spar; and polar winding the trailing edge. The inner torque tube is wound over a rubber bag supported by a metal mandrel. After winding, the metal part is removed so that the torque tube wrap can be pressurized and cured in a matched-die mold along with the spar caps. This assembly is then covered with helical windings and remolded to form the complete spar assembly. The formation of the rest of the blade, including the afterbody airfoil (±45° fibers on a honeycomb core), is indicated in Fig. 6.10.

Braiding

Braiding is a form of filament winding in which a number of filament spools, mounted on a rotating ring, produce a woven layer on a central mandrel as it moves through the ring. Although the mechanical properties of braided composites are comparable only with woven cloth (apart from the interlaminar shear strength, which is very high due to the extensive interweaving of the fibers), the process offers a number of advantages: (1) noncircular shapes (such as box sections) and complex shapes (such as forks and bends in pipes), including section changes, can be braided; (2) deposition of the composite is rapid; (3) continuous forms can be produced as in pultrusion, but with angular as well as longitudinal reinforcement; and (4) the braided material can be remolded. Glass, aramid, and graphite fibers can be braided; however, graphite fibers, even when sized, are difficult to handle without excessive fiber damage.

6.5 PULTRUSION

Continuous uniform sections such as rods, tubes, and beams (I sections, etc.) can be made by pultrusion.[2] In this process, schematically illustrated in Fig. 6.11, continuous fibers from individual spools are pulled through a resin bath, combined, and then pulled through a forming die. The die,

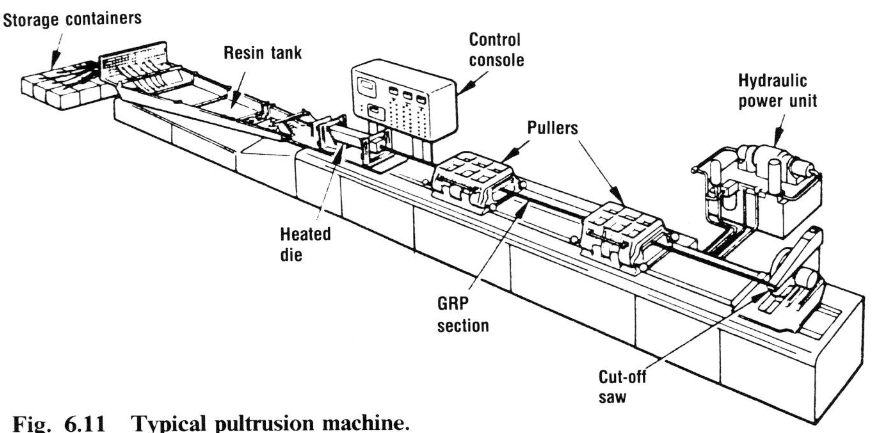

Fig. 6.11 Typical pultrusion machine.

usually of chromium-plated steel, forms the section shape and squeezes out the excess resin; in some cases, the die is heated and may then partially or finally cure the composite. Usually, however, an oven (or alternatively a microwave unit) is used to effect final cure.

Structures with very high fiber volume fractions (approximately 70%) can be produced, giving excellent unidirectional strength and stiffness. However, as may be expected, the transverse strength is low. This limitation can be reduced by incorporating glass cloth or mat into the structure.

One of the major problems with pultrusion is that surface damage occurs due to frictional forces in the die—particularly if the dies do not have a sufficiently gradual section decrease. Friction causes pieces of semicured resin to shear off and attach to the die surface (where they can cure) and then score the surface of the material following. This problem can be reduced by periodically stopping the run and allowing the separate resin particles to attach themselves to the material in the die. When the run is continued, the die is swept clean.

Glass, aramid, and graphite fibers can be successfully pultruded. A resin size is generally used on graphite fibers to avoid handling damage. Pultruded glass fiber-reinforced polyester sections are readily available off the shelf.

Both polyester and epoxy resins may be used. Sometimes, a small amount of an internal release agent such as an alcohol phosphate is added to the resin to reduce friction drag during pultruding. However, this addition can adversely affect the properties of the composite.

6.6 MACHINING

Graphite/epoxy materials pose no major machining problems. However, aramid/epoxy causes difficulties due to its inherent toughness and, thus, specially designed tools are required. Boron/epoxy is very difficult to machine as a result of the extreme hardness of the boron fibers, which is close to that of diamond. Conventional machining operations require diamond-impregnated tools or, alternatively, ultrasonic procedures with diamond paste.

Machining and drilling of graphite/epoxy[2] can be satisfactorily performed with tungsten-carbide tipped tools. The main difficulty with drilling is that the composite can be delaminated fairly readily if it is not firmly supported during drilling (e.g., by sandwiching it between scrap material) and if the rate of drill feed is too high. A particular problem is the tendency for local tearing during drill breakthrough. This difficulty can be greatly reduced by bonding a layer of film adhesive to the exit face of the laminate.

Milling can be performed with conventional tungsten-carbide end mills. Damage to the laminate is avoided if the mill passes are made in the fiber direction, since this minimizes fiber tear-out. Surface grinding can best be performed with silicon-carbide wheels. Cutting and slitting are best performed with diamond-impregnated tools.

The advanced cutting procedures employing lasers, water jets, or ultrasonic techniques are effective with all of the advanced composites.

References

[1]Sonneborn, R. H., Dietz, A. G., and Heysen, A. S., *Fiberglass Reinforced Plastics*, Reinhold Publishing Co., New York, 1954.

[2]Langley, M. (ed.), *Carbon Fibres in Engineering*, McGraw-Hill Book Co., London, 1973, Chap. 3.

[3]Lackman, C. M., O'Brien, W. L., and Loyd, M. S., "Advanced Composites Integral Structures Meet the Challenge of Future Aircraft Systems," *Fibrous Composites in Structural Design*, edited by E. M. Lenoe, D. W. Oplinger, and J. J. Burke, Plenum Press, New York, 1980, pp. 125–144.

[4]May, C. A. (ed.), "Instrumental Cure Monitoring—State of Art," *Proceedings of 22nd National SAMPE Symposium*, Vol. 22, 1977, pp. 618–662.

[5]Carpenter, J. F. and Bartels, T. T., "Characterization and Control of Composite Pre-pregs and Adhesives," *SAMPE Quarterly*, Vol. 7, 1976, pp. 1–7.

[6]Rosato, D. V. and Grove, C. S., *Filament Winding*, Interscience, New York, 1964.

[7]Ainsworth, L., "The State of Filament Winding," *Composites*, Vol. 2, 1971, p. 14.

[8]Covington, C. E. and Baumgardner, P. S., "Design and Production of Fiberglass Helicopter Rotor Blades," *Fibrous Composites in Structural Design*, by E. M. Lenoe, D. W. Oplinger, and J. J. Burke, Plenum Press, New York, 1980, pp. 497–515.

7. STRUCTURAL MECHANICS OF FIBER COMPOSITES

7.1 INTRODUCTION

In this chapter, the basic theory needed for the determination of the stresses, strains, and deformations in fiber composite structures is outlined. Attention is concentrated on structures made in the form of laminates because that is the way composite materials are generally utilized.

From the viewpoint of structural mechanics, the novel features of composites, compared with conventional structural materials such as metals, are their marked anisotropy and, when used as laminates, their macroscopically heterogeneous nature. However, it should be remarked that there is one classical structural material, namely, wood, that is also both anisotropic and, when used in the form of plywood, macroscopically heterogeneous. In fact, much of the theory that is needed in the analysis of composite structures is simply an extension of the theory already used for wooden structures.

There is a close analogy between the steps in developing laminate theory and the steps in fabricating a laminate. The building block both for theory and fabrication is the "single ply" (also referred to as the "basic ply," "monoply," "monolayer," or "lamina"). This is a thin layer of the material (a typical thickness being around 0.125 mm for graphite/epoxy) in which all of the fibers are aligned parallel to one another.* The starting point for the theory is the stress-strain law for the single ply referred to its axes of material symmetry. In constructing a laminate, each ply is laid up so that its fibers make some prescribed angle with a reference axis fixed in the laminate. All later calculations are made using axes fixed in the laminate (the "structural axes") so that it is necessary to transform the stress-strain law to these axes. Finally, the individual plies are bonded into a structural entity; thus, it is necessary to combine the stress-strain laws for the individual plies into a single stress-strain law for the laminate. Since the designer can select his own lay-up pattern and since the laminate stress-strain

*There is some interest in making composite aircraft structures from graphite/epoxy woven cloth rather than from unidirectional plies, although, generally speaking, more efficient structures can be obtained using the latter form of material. If cloth is used, then some of the statements made here need to be modified since, of course, cloth has fibers running in two perpendicular directions. However, the theoretical formulas developed in this chapter can be readily adapted to apply to a laminate made from cloth; the main difference is in the values of the material constants used in the formulas.

law will depend on that pattern, it follows that the designer can "design the material" (as well as the structure).

For more detailed discussions of the topics covered in this chapter, see, for example, Refs. 1–7. For background material on the theory of anisotropic elasticity, see, Refs. 8–10.

7.2 STRESS-STRAIN LAW FOR SINGLE PLY IN MATERIAL AXES: UNIDIRECTIONAL LAMINATES

Consider a rectangular element of a single ply with the sides of the element parallel and perpendicular to the fiber direction (Fig. 7.1). Clearly, the direction of the fibers defines a preferred direction in the material; it is thus natural to introduce a Cartesian set of "material axes" 0-1, 2, 3 with the 1 axis in the fiber direction, the 2 axis perpendicular to the fibers and in the ply plane, and the 3 axis perpendicular to the plane of the ply. Here, interest is in the behavior of the ply when subjected to stresses acting in its plane, i.e., under plane stress conditions; these stresses (also referred to the material axes) will be denoted by σ_1, σ_2, and τ_{12} and the associated strains by ε_1, ε_2, and γ_{12}. (Note that, in composite mechanics, it is standard practice to work with "engineering" rather than "tensor" shear strains.) Although a single ply is highly anisotropic, it is intuitively evident that the coordinate planes 012, 023, and 031 are those of material symmetry, there being a mirror image symmetry about these planes. A material having three mutually orthogonal planes of symmetry is known as "orthotropic." As is shown, for example, by Tsai,[11] the stress-strain law for an orthotropic material under plane stress conditions, referred to the material axes necessarily has the following form:

$$\begin{bmatrix} \varepsilon_1 \\ \varepsilon_2 \\ \gamma_{12} \end{bmatrix} = \begin{bmatrix} 1/E_1 & -\nu_{21}/E_2 & 0 \\ -\nu_{12}/E_1 & 1/E_2 & 0 \\ 0 & 0 & 1/G_{12} \end{bmatrix} \begin{bmatrix} \sigma_1 \\ \sigma_2 \\ \tau_{12} \end{bmatrix} \tag{7.1}$$

where: E_1, E_2 = Young's moduli in the 1 and 2 directions, respectively,

ν_{12} = Poisson's ratio governing the contraction in the 2 direction for a tension in the 1 direction

ν_{21} = Poisson's ratio governing the contraction in the 1 direction for a tension in the 2 direction

G_{12} = (in-plane) shear modulus

There are five material constants in Eq. (7.1), but only four of these are independent because of the following symmetry relation:

$$\nu_{12}/E_1 = \nu_{21}/E_2 \tag{7.2}$$

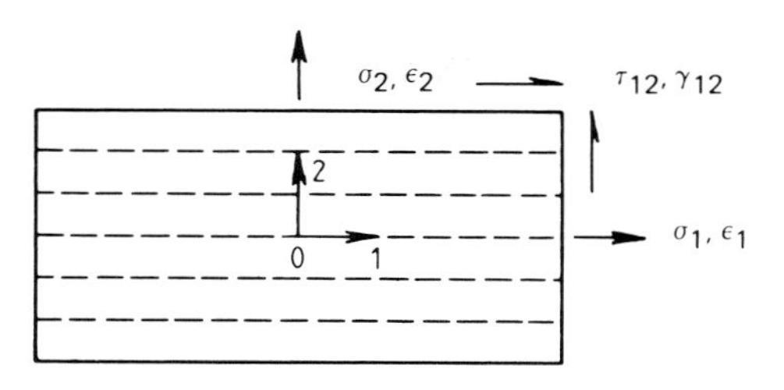

Fig. 7.1 Material axes for a single ply.

For composites of the type being considered here, it should always be borne in mind that E_1 is much larger than either E_2 or G_{12} because the former is a "fiber-dominated" property, while the latter are "matrix-dominated."

The above equations are all related to a single ply; but, since the ply thickness does not enter into the calculations, they also apply to a "unidirectional laminate" that is simply a laminate in which the fiber direction is the same in all of the plies. In fact, most of the material constants for a single ply are obtained from specimen tests on unidirectional laminates, a single ply being itself too thin to test conveniently.

For much of the later analysis, it is more convenient to deal with the inverse form of Eq. (7.1), namely,

$$\begin{bmatrix} \sigma_1 \\ \sigma_2 \\ \tau_{12} \end{bmatrix} = \begin{bmatrix} Q_{11}(0) & Q_{12}(0) & 0 \\ Q_{12}(0) & Q_{22}(0) & 0 \\ 0 & 0 & Q_{66}(0) \end{bmatrix} \begin{bmatrix} \varepsilon_1 \\ \varepsilon_2 \\ \gamma_{12} \end{bmatrix} \tag{7.3}$$

where the $Q_{ij}(0)$, commonly termed the reduced stiffness coefficients, are given by

$$Q_{11}(0) = E_1/(1 - \nu_{12}\nu_{21}) \qquad Q_{22}(0) = E_2/(1 - \nu_{12}\nu_{21})$$

$$Q_{12}(0) = \nu_{21}E_1/(1 - \nu_{12}\nu_{21}) \qquad Q_{66}(0) = G_{12} \tag{7.4}$$

It is conventional in composite mechanics to use the above subscript notation for Q, the point of which becomes evident only when three-dimensional anisotropic problems are encountered.

7.3 STRESS-STRAIN LAW FOR SINGLE PLY IN STRUCTURAL AXES: OFF-AXIS LAMINATES

As already noted, when a ply is incorporated in a laminate, its fibers will make some prescribed angle θ with a reference axis fixed in the laminate. Let this be the x axis and note that the angle θ is measured from the x axis to the 1 axis and is positive in the counterclockwise direction; the y axis is perpendicular to the x axis and in the plane of the ply. (See Fig. 7.2.) All

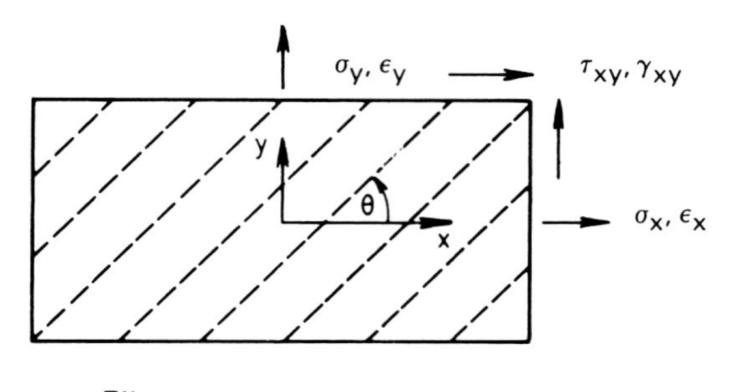

Fig. 7.2 Structural axes for a single ply.

subsequent calculations are made using the xy, or "structural," axes, so it is necessary to transform the stress-strain law from the material axes to the structural axes. If the stresses in the structural axes are denoted by σ_x, σ_y, and τ_{xy}, then these are related to the stresses referred to the material axes by the usual transformation equations,

$$\begin{bmatrix} \sigma_x \\ \sigma_y \\ \tau_{xy} \end{bmatrix} = \begin{bmatrix} c^2 & s^2 & -2cs \\ s^2 & c^2 & 2cs \\ cs & -cs & c^2 - s^2 \end{bmatrix} \begin{bmatrix} \sigma_1 \\ \sigma_2 \\ \tau_{12} \end{bmatrix} \tag{7.5}$$

where c denotes $\cos\theta$ and s $\sin\theta$. Also, the strains in the material axes are related to those in the structural axes, namely, ε_x, ε_y, and γ_{xy}, by what is essentially the inverse transformation,

$$\begin{bmatrix} \varepsilon_1 \\ \varepsilon_2 \\ \gamma_{12} \end{bmatrix} = \begin{bmatrix} c^2 & s^2 & cs \\ s^2 & c^2 & -cs \\ -2cs & 2cs & c^2 - s^2 \end{bmatrix} \begin{bmatrix} \varepsilon_x \\ \varepsilon_y \\ \gamma_{xy} \end{bmatrix} \tag{7.6}$$

Now, in Eqs. (7.5), substitute for σ_1, σ_2, and τ_{12} their values as given by Eqs. (7.3); then, in the resultant equations, substitute for ε_1, ε_2, and γ_{12} their values as given by Eqs. (7.6). After some routine manipulations, it is found that the stress-strain law in the structural axes has the form

$$\begin{bmatrix} \sigma_x \\ \sigma_y \\ \tau_{xy} \end{bmatrix} = \begin{bmatrix} Q_{11}(\theta) & Q_{12}(\theta) & Q_{16}(\theta) \\ Q_{12}(\theta) & Q_{22}(\theta) & Q_{26}(\theta) \\ Q_{16}(\theta) & Q_{26}(\theta) & Q_{66}(\theta) \end{bmatrix} \begin{bmatrix} \varepsilon_x \\ \varepsilon_y \\ \gamma_{xy} \end{bmatrix} \tag{7.7}$$

where the $Q_{ij}(\theta)$ are related to the $Q_{ij}(0)$ by the following equations:

$$\begin{bmatrix} Q_{11}(\theta) \\ Q_{12}(\theta) \\ Q_{22}(\theta) \\ Q_{16}(\theta) \\ Q_{26}(\theta) \\ Q_{66}(\theta) \end{bmatrix} = \begin{bmatrix} c^4 & 2c^2s^2 & s^4 & 4c^2s^2 \\ c^2s^2 & c^4+s^4 & c^2s^2 & -4c^2s^2 \\ s^4 & 2c^2s^2 & c^4 & 4c^2s^2 \\ c^3s & -cs(c^2-s^2) & -cs^3 & -2cs(c^2-s^2) \\ cs^3 & cs(c^2-s^2) & -c^3s & 2cs(c^2-s^2) \\ c^2s^2 & -2c^2s^2 & c^2s^2 & (c^2-s^2)^2 \end{bmatrix} \times \begin{bmatrix} Q_{11}(0) \\ Q_{12}(0) \\ Q_{22}(0) \\ Q_{66}(0) \end{bmatrix} \tag{7.8}$$

Observe that, in Eqs. (7.7), the direct stresses depend on the shear strains (as well as the direct strains) and the shear stress depends on the direct strains (as well as the shear strain). This complication arises because, for nonzero θ, the structural axes are not axes of material symmetry and, with respect to these axes, the material is not orthotropic; it is evident that the absence of orthotropy leads to the presence of the Q_{16} and Q_{26} terms in Eqs. (7.7). Also, for future reference, note that the expressions for $Q_{11}(\theta)$, $Q_{12}(\theta)$, $Q_{22}(\theta)$, and $Q_{66}(\theta)$ contain only even powers of $\sin\theta$ and so these quantities are unchanged when θ is replaced by $-\theta$. On the other hand, the expressions for $Q_{16}(\theta)$ and $Q_{26}(\theta)$ contain odd powers of $\sin\theta$ and so they change sign when θ is replaced by $-\theta$.

Analogously to the previous section, the above discussion has been related to a single ply, but it is equally valid for a laminate in which the fiber direction is the same in all plies; a unidirectional laminate in which the fiber direction makes a nonzero angle with the x structural axis is known as an "off-axis" laminate and is sometimes used for test purposes. Formulas for the elastic moduli of an off-axis laminate can be obtained by a procedure analogous to that used in deriving Eqs. (7.7). Using Eqs. (7.1), with the inverse forms of Eqs. (7.5) and (7.6), leads to the inverse form of Eqs. (7.7), i.e., with the strains expressed in terms of the stresses; from this result, the moduli can be written. Details can be found in most of the standard texts, e.g., p. 54 of Ref. 1. Only the result for the Young's modulus in the x direction, E_x, will be cited here,

$$1/E_x = (1/E_1)c^4 + (1/G_{12} - 2\nu_{12}/E_1)c^2s^2 + (1/E_2)s^4 \tag{7.9}$$

The variation of E_x with θ for the case of a graphite/epoxy off-axis laminate is shown in Fig. 7.3. The material constants of the single ply were taken to be

$$E_1 = 137.8 \text{ GPa} \qquad E_2 = 11.71 \text{ GPa} \qquad G_{12} = 5.51 \text{ GPa}$$

$$\nu_{12} = 0.25 \qquad \nu_{21} = 0.0213$$

It can be seen that the modulus initially decreases quite rapidly as the off-axis angle increases from 0°; this indicates the importance of the precise alignment of fibers in a laminate.

7.4 PLANE STRESS PROBLEMS FOR SYMMETRIC LAMINATES

General

One of the commonest structural forms for composites is a laminated sheet loaded in its own plane, i.e., under plane stress conditions. In order that out-of-plane bending will not occur, such a laminate is always made with a lay-up that is symmetric about the midthickness plane. Just to illustrate the type of symmetry meant, consider an eight-ply laminate comprising four plies which are to be oriented at 0° to the reference (x) axis, two plies at +45°, and two plies at −45°. An example of a symmetric laminate would be one with the following ply sequence:

$$0°/0°/+45°/-45°/-45°/+45°/0°/0°$$

On the other hand, an example of an unsymmetric arrangement of the same plies would be

$$0°/0°/0°/0°/+45°/-45°/+45°/-45°$$

These two cases are shown in Fig. 7.4 where z denotes the coordinate in the thickness direction.

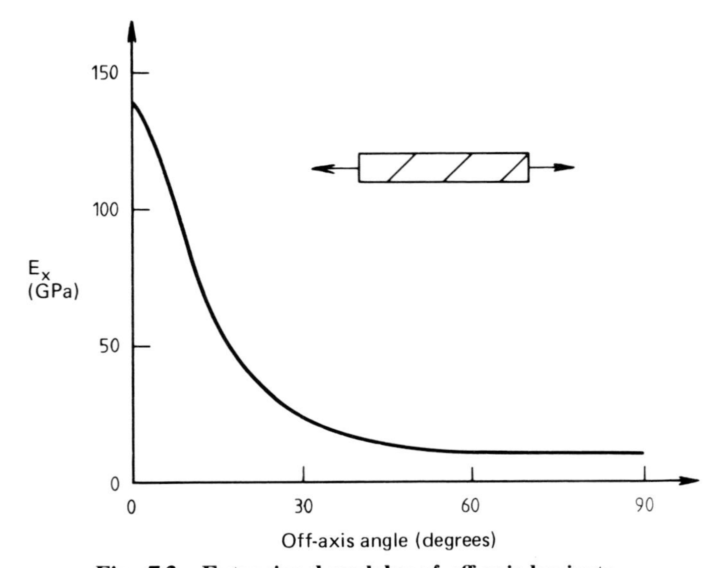

Fig. 7.3 Extensional modulus of off-axis laminate.

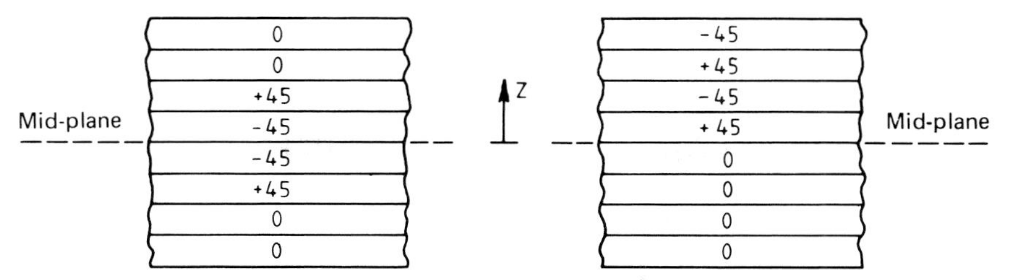

Fig. 7.4 Symmetric (left) and nonsymmetric (right) eight-ply laminates.

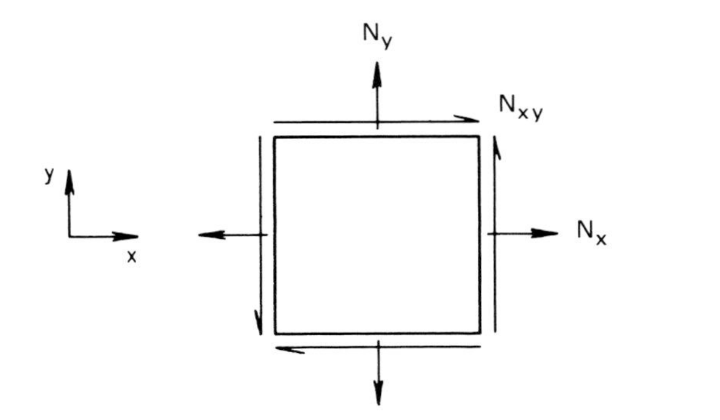

Fig. 7.5 Stress resultants.

Laminate Stiffness Matrix

Consider now a laminate comprising n plies and denote the angle between the fiber direction in the kth ply and the x structural axis by θ_k (with the same convention as described in Sec. 7.3); subject only to the symmetry requirement, the ply orientation is arbitrary. It is assumed that, when the plies are molded into the laminate, a rigid bond (of infinitesimal thickness) is formed between adjacent plies; as a consequence of this assumption, it follows that under plane stress conditions the strains are the same at all points on a line through the thickness (i.e., that they are independent of z). Denoting these strains by ε_x, ε_y, and γ_{xy}, it then follows from Eqs. (7.7) that the stresses in the kth ply will be given by

$$\sigma_x(k) = Q_{11}(\theta_k)\varepsilon_x + Q_{12}(\theta_k)\varepsilon_y + Q_{16}(\theta_k)\gamma_{xy}$$

$$\sigma_y(k) = Q_{12}(\theta_k)\varepsilon_x + Q_{22}(\theta_k)\varepsilon_y + Q_{26}(\theta_k)\gamma_{xy}$$

$$\tau_{xy}(k) = Q_{16}(\theta_k)\varepsilon_x + Q_{26}(\theta_k)\varepsilon_y + Q_{66}(\theta_k)\gamma_{xy} \tag{7.10}$$

If the laminate thickness is denoted by h and assuming all plies are of the same thickness (which is the usual situation), then the thickness of an individual ply is simply h/n. Now consider an element of the laminate with sides of unit length parallel to the x and y axes. The forces on this element will be denoted by N_x, N_y, and N_{xy} (Fig. 7.5); the N are generally termed "stress resultants" and have the dimension "force/length." Elementary equilibrium considerations give

$$N_x = \left(\frac{h}{n}\right) \sum_{k=1}^{n} \sigma_x(k), \qquad N_y = \left(\frac{h}{n}\right) \sum_{k=1}^{n} \sigma_y(k), \qquad N_{xy} = \left(\frac{h}{n}\right) \sum_{k=1}^{n} \tau_{xy}(k) \tag{7.11}$$

Substituting from Eqs. (7.10) into (7.11) and remembering that the strains are the same in all plies, the following result is readily obtained:

$$\begin{aligned} N_x &= A_{11}\varepsilon_x + A_{12}\varepsilon_y + A_{16}\gamma_{xy} \\ N_y &= A_{12}\varepsilon_x + A_{22}\varepsilon_y + A_{26}\gamma_{xy} \\ N_{xy} &= A_{16}\varepsilon_x + A_{26}\varepsilon_y + A_{66}\gamma_{xy} \end{aligned} \tag{7.12}$$

where

$$A_{ij} = \left(\frac{h}{n}\right) \sum_{k=1}^{n} Q_{ij}(\theta_k) \tag{7.13}$$

The quantities A_{ij} are the terms of the laminate "in-plane stiffness matrix"; given the single-ply moduli and the laminate lay-up details, they can be calculated routinely by using Eqs. (7.4), (7.8), and (7.13). Equations (7.12) are generally taken as the starting point for any laminate structural analysis.

Laminate Stress-Strain Law

As was just implied, it seems to be the current fashion in laminate mechanics to work in terms of the stress resultants, rather than the stresses; however, for some purposes, the latter are more convenient. From the stress resultants, the average stresses (averaged through the thickness of the laminate) are easily obtained; writing these stresses simply as σ_x, σ_y, and τ_{xy}, then

$$\sigma_x = N_x/h, \qquad \sigma_y = N_y/h, \qquad \tau_{xy} = N_{xy}/h \tag{7.14}$$

Hence, in terms of these average stresses, the stress-strain law for the laminate becomes

$$\begin{aligned} \sigma_x &= A^*_{11}\varepsilon_x + A^*_{12}\varepsilon_y + A^*_{16}\gamma_{xy} \\ \sigma_y &= A^*_{12}\varepsilon_x + A^*_{22}\varepsilon_y + A^*_{26}\gamma_{xy} \\ \tau_{xy} &= A^*_{16}\varepsilon_x + A^*_{26}\varepsilon_y + A^*_{66}\gamma_{xy} \end{aligned} \tag{7.15}$$

where

$$A^*_{ij} = A_{ij}/h = \left[\sum_{k=1}^{n} Q_{ij}(\theta_k) \right] / n \tag{7.16}$$

In some cases, Eqs. (7.15) are more convenient than are Eqs. (7.12).

Orthotropic Laminates

An orthotropic laminate, having the structural axes as the axes of orthotropy, is one for which $A_{16} = A_{26} = 0$; clearly, this implies that

$$\sum_{k=1}^{n} Q_{16}(\theta_k) = 0, \qquad \sum_{k=1}^{n} Q_{26}(\theta_k) = 0 \tag{7.17}$$

Thus, the stress-strain law for an orthotropic laminate reduces to

$$\sigma_x = A^*_{11}\varepsilon_x + A^*_{12}\varepsilon_y$$

$$\sigma_y = A^*_{12}\varepsilon_x + A^*_{22}\varepsilon_y$$

$$\tau_{xy} = A^*_{66}\gamma_{xy} \tag{7.18}$$

The coupling between the direct stresses and the shear strains and between the shear stresses and the direct strains, which is present for a general laminate, disappears for an orthotropic laminate. Most laminates currently in use are orthotropic.

It can be readily seen that the following laminates will be orthotropic:

(1) Those consisting only of plies for which $\theta = 0°$ or $90°$; here it follows from Eqs. (7.8) that in either case $Q_{16}(\theta) = Q_{26}(\theta) = 0$.

(2) Those constructed such that for each ply oriented at an angle θ, there is another ply oriented at an angle $-\theta$; since, as already noted in Sec. 7.3,

$$Q_{16}(-\theta) = -Q_{16}(\theta), \qquad Q_{26}(-\theta) = -Q_{26}(\theta)$$

there is then a cancellation of all paired terms in the summation of Eq. (7.17).

(3) Those consisting only of 0°, 90°, and matched pairs of $\pm\theta$ plies are also, of course, orthotropic.

An example of an orthotropic laminate would be one with the following ply pattern:

$$0°/+30°/-30°/-30°/+30°/0°$$

On the other hand, the following laminate (while still symmetric) would not be orthotropic:

$$0°/+30°/90°/90°/+30°/0°$$

Moduli of Orthotropic Laminates

Expressions for the moduli of orthotropic laminates can easily be obtained by solving Eqs. (7.18) for simple loadings. For example, on setting $\sigma_y = \tau_{xy} = 0$, Young's modulus in the x direction, E_x, and Poisson's ratio ν_{xy} governing the contraction in the y direction for a stress in the x direction are then given by

$$E_x = \sigma_x/\varepsilon_x, \qquad \nu_{xy} = -\varepsilon_y/\varepsilon_x$$

Proceeding in this way it is found that

$$E_x = A_{11}^* - A_{12}^{*2}/A_{22}^* \qquad E_y = A_{22}^* - A_{12}^{*2}/A_{11}^*$$

$$\nu_{xy} = A_{12}^*/A_{22}^* \qquad \nu_{yx} = A_{12}^*/A_{11}^* \qquad G_{xy} = A_{66}^* \tag{7.19}$$

(It is possible to derive analogous, but more complicated, formulas for the moduli of nonorthotropic laminates by solving Eqs. (7.15) for simple loadings; see Ref. 12.

As illustrative examples of the above theory consider a family of 24-ply laminates, symmetrical and orthotropic, and all made of the same material but with varying numbers of 0° and ±45° plies. (For the present purposes, the precise ordering of the plies is immaterial as long as the symmetry requirement is maintained; however, to ensure orthotropy, there must be the same number of +45° as −45° plies). The single-ply modulus data (representative of a graphite/epoxy) are

$$E_1 = 137.8 \text{ GPa} \qquad E_2 = 11.71 \text{ GPa} \qquad G_{12} = 5.51 \text{ GPa}$$

$$\nu_{12} = 0.25 \qquad \nu_{21} = 0.0213$$

The lay-ups considered are shown in Table 7.1. Just to recapitulate, the steps in the calculation are as follows:

(1) Calculate the $Q_{ij}(0)$ from Eqs. (7.4).

(2) For each of the ply orientations involved—here $\theta = 0°$, $+45°$, and

Table 7.1 Moduli for Family of 24-ply 0° / ± 45° Laminates

Lay-Up							
No. 0° Plies	No. +45° Plies	No. −45° Plies	E_x, GPa	E_y, GPa	G_{xy}, GPa	ν_{xy}	ν_{yx}
24	0	0	137.8	11.7	5.51	0.250	0.021
16	4	4	99.6	21.1	15.7	0.579	0.123
12	6	6	79.7	24.5	20.8	0.648	0.199
8	8	8	59.7	26.5	25.9	0.694	0.308
0	12	12	19.3	19.3	36.1	0.753	0.753

$-45°$—calculate the $Q_{ij}(\theta)$ from Eqs. (7.8). (Of course, here the $Q_{ij}(0)$ have already been obtained in step 1.)

(3) Calculate the A^*_{ij} from Eq. (7.16); in the present case, Eq. (7.16) becomes

$$A^*_{ij} = [n_1 Q_{ij}(0) + n_2 Q_{ij}(+45) + n_3 Q_{ij}(-45)]/24$$

where n_1 is the number of 0° plies, n_2 of +45° plies, and n_3 of −45° plies.

(4) Calculate the moduli from Eqs. (7.19).

The results of the calculations are shown in Table 7.1.

The results in Table 7.1 have been presented primarily to exemplify the preceding theory; however, they also demonstrate some features that are important in design. The stiffness of a composite is overwhelmingly resident in the extensional stiffness of its fibers; hence, at least for simple loadings, if maximum stiffness is required, a laminate is constructed so that the fibers are aligned in the principal stress directions. Thus, for a member under uniaxial tension, a laminate comprising basically all 0° plies would be chosen, i.e., with all fibers aligned parallel to the tension direction. As can be seen from Table 7.1, E_x decreases as the number of 0° plies decreases. On the other hand, consider a rectangular panel under shear, the sides of the panel being parallel to the structural axes (Fig. 7.6). The principal stresses here are an equal tension and compression, oriented at +45° and −45° to the x axis. Thus, maximum shear stiffness can be expected to be obtained using a laminate comprising equal numbers of +45° and −45° plies; this is reflected in the high shear modulus G_{xy} for the all ±45° laminate of Table 7.1. (For comparison, recall that Young's modulus and the shear modulus for a typical aluminum alloy are of the order of 72 and 27 GPa, respectively, and that the specific gravity of graphite/epoxy is about 60% that of aluminum.)

It should also be observed that, while for an isotropic material, Poisson's ratio cannot exceed 0.5, this is not the case for an anisotropic material.

Quasi-isotropic Laminates

It is possible to construct laminates that are isotropic as regards their in-plane elastic properties, i.e., that have the same Young's modulus E and

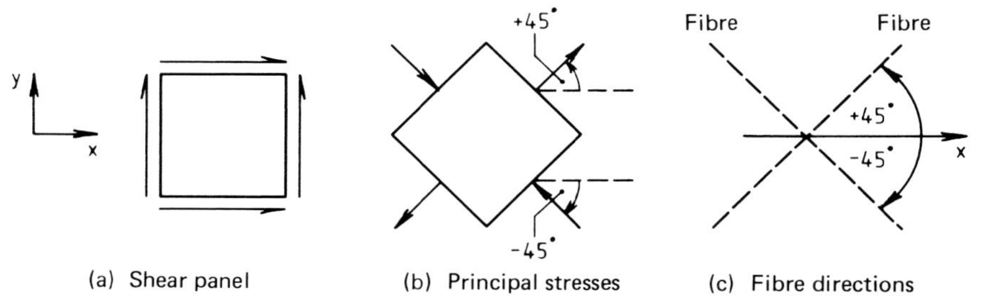

Fig. 7.6 Fiber orientations for a shear panel.

same Poisson's ratio ν in all in-plane directions and for which the shear modulus is given by $G = E/2(1 + \nu)$. One way of achieving this is to adopt a lay-up having an equal number of plies oriented parallel to the sides of an equilateral triangle. For example, a quasi-isotropic 24-ply laminate could be made with 8 plies oriented at each of 0°, +60°, and −60°. Using the same materials data (and theory) as were used in deriving Table 7.1, it will be found that such a laminate has the following moduli:

$$E = 54.3 \text{ GPa}, \qquad G = 20.8 \text{ GPa}, \qquad \nu = 0.305$$

Another way of achieving a quasi-isotropic laminate is to use equal number of plies oriented at 0°, +45°, −45°, and 90°, i.e., an alternative lay-up that would also give a quasi-isotropic 24-ply laminate (with, incidentally, the same values for the elastic constants as were just cited) could be made with 6 plies at each of 0°, +45°, −45°, and 90°.

The term "quasi-isotropic" is used because, of course, such laminates have different properties in the out-of-plane direction. However, it is not usual practice to work with quasi-isotropic laminates; efficient design with composites generally requires that advantage be taken of their inherent anisotropy.

Stress Analysis of Orthotropic Laminates

The determination of the stresses, strains, and deformations experienced by symmetric laminates under plane stress loadings is carried out by procedures that are analogous to those used for isotropic materials. The laminate is treated as a homogeneous membrane having stiffness properties determined as described above. It should be noted, though, that while the strains and deformations so determined are the actual strains and deformations (within the limit of the assumptions), the stresses are only the average values over the laminate thickness.

If an analytical procedure is used, then generally a stress function F is introduced, this being related to the (average) stresses by

$$\sigma_x = \frac{\partial^2 F}{\partial y^2}, \qquad \sigma_y = \frac{\partial^2 F}{\partial x^2}, \qquad \tau_{xy} = \frac{-\partial^2 F}{\partial x \partial y} \tag{7.20}$$

It can be shown that F satisfies the following partial differential equation:

$$\left(\frac{1}{E_y}\right)\frac{\partial^4 F}{\partial x^4} + \left(\frac{1}{G_{xy}} - \frac{2\nu_{xy}}{E_x}\right)\frac{\partial^4 F}{\partial x^2 \partial y^2} + \left(\frac{1}{E_x}\right)\frac{\partial^4 F}{\partial y^4} = 0 \tag{7.21}$$

Solutions of Eq. (7.21) for several problems of interest can be found in Ref. 10.

These days, most structural analyses are performed using finite element methods and many general-purpose finite element programs contain orthotropic membrane elements in their library. Once the laminate moduli or stiffness coefficients (depending on whether the program is in terms of stresses or stress resultants) are determined as above, these are used as input data for calculating the element stiffness matrix; the rest of the analysis proceeds as in the isotropic case.

As has already been emphasized, the stresses obtained from the above procedures are only the average stresses; however, to determine the actual stresses in the individual plies, it is necessary only to substitute the calculated values of the strains in Eqs. (7.10). An elementary example may clarify this. Consider a rectangular strip under uniaxial tension (Fig. 7.7) made of the 24-ply laminate considered earlier that had 12 plies at 0° and 12 plies at ±45°; suppose the applied stress is $\sigma_x = 68.9$ MPa. The average stress here is uniform in the xy plane and given by

$$\sigma_x = 68.9 \text{ MPa}, \qquad \sigma_y = \tau_{xy} = 0$$

Using the values of E_x and ν_{xy} given in Table 7.1, it follows that the associated strains are

$$\varepsilon_x = 0.864 \times 10^{-3}, \qquad \varepsilon_y = -0.560 \times 10^{-3}, \qquad \gamma_{xy} = 0$$

The stresses in the individual plies can now be obtained by substituting these values into Eqs. (7.10), with the appropriate values of the $Q_{ij}(\theta)$. The results of doing this are shown in Table 7.2.

Thus, the actual stress distribution is very different from the average one. In particular, note that transverse direct stresses and shear stresses are developed, even though no such stresses are applied; naturally, these stresses are self-equilibrating over the thickness. It follows that there is some

Table 7.2 Stresses in Individual Plies of 24-Ply Laminate (12 at 0°, 12 at ±45°) under Uniaxial Stress of 68.9 MPa

θ,°	$\sigma_x(\theta)$, MPa	$\sigma_y(\theta)$, MPa	$\tau_{xy}(\theta)$, MPa
0	118.1	−4.0	0
+45	19.7	4.0	9.7
−45	19.7	4.0	−9.7

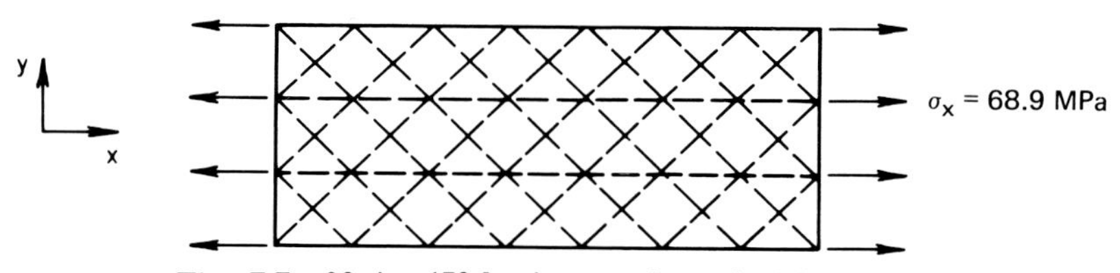

Fig. 7.7 0°/±45° laminate under uniaxial tension.

"boundary layer" around the edges of the strip where there is a rapid transition from the actual stress boundary values (namely, zero on the longitudinal edges) to the above calculated values. This boundary layer would be expected to extend in from the edges a distance of the order of the laminate thickness (from St. Venant's principle). In the boundary layer, the simple laminate theory presented above is not applicable and a three-dimensional analysis is required (see p. 191 of Ref. 5). The matter is of more than academic interest since faults, such as delaminations, are prone to originate at the free edges of laminates because of the above effect.

Finally, when discussing allowable design values for composite structures, it is usual to cite values of strain, rather than stress; clearly, strain is the more meaningful quantity for a laminate.

Stress Concentration Around Holes in Orthotropic Laminates

Several analytical solutions for the stresses around holes in (symmetric) orthotropic laminates are cited in Ref. 7; details of the derivations of these are given in Ref. 10. It turns out that the value of the stress concentration factor (SCF) depends markedly on the relative values of the various moduli. This can be illustrated by considering the case of a circular hole in an infinite sheet under a uniaxial tension in the x direction (Fig. 7.8); here the stress concentration factor at point A in Fig. 7.8 ($\alpha = 90°$) is given by the following formula:

$$\mathrm{SCF} = 1 + \left\{2\left[(E_x/E_y)^{\frac{1}{2}} - \nu_{xy}\right] + E_x/G_{xy}\right\}^{\frac{1}{2}} \tag{7.22}$$

The stress concentration factors for the laminates of Table 7.1 have been calculated from this formula and the results are shown in Table 7.3.

For comparison, the SCF for an isotropic material is 3. As can be seen, when there is a high degree of anisotropy (e.g., an all 0° laminate), SCFs well in excess of that can be obtained. It should also be pointed out that, as the laminate pattern changes, not only does the value of the SCF change,

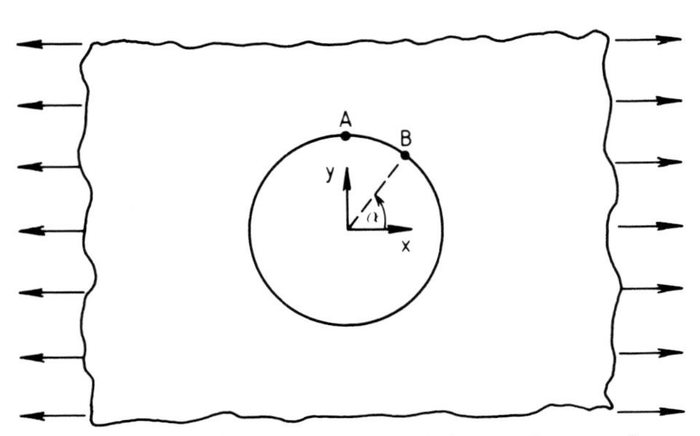

Fig. 7.8 Circular hole in infinite tension panel.

but the point where the SCF attains its maximum value can also change. Whereas for the first four laminates of Table 7.3, the maximum SCF does occur at point A, for the remaining laminate the maximum occurs at a point such as B in Fig. 7.8.

Laminate Codes

Although the precise ordering of the plies of a laminate has not been of concern in the considerations of this section, generally there will be other factors that will determine such an ordering. In any case, when a laminate is being called up for manufacture, the associated engineering drawing should list the orientation of each ply. When referring to laminates in a text, some sort of abbreviated notation is necessary to specify the pattern. It is easiest to describe the code normally used by some examples.

Example 1

Consider an eight-ply laminate with the following (symmetric) lay-up

$$0°/0°/+45°/-45°/-45°/+45°/0°/0°$$

This is written in code form as

$$[0_2/\pm 45]_s$$

Note that

(1) Only half the plies in a symmetric laminate are listed, the symmetry being implied by the s outside the brackets.

(2) The degree signs are omitted from the angles.

(3) In a + and − combination, the upper sign is read first.

Example 2

Consider a 50-ply laminate containing repetitions of the ply sequence

$$0°/0°/+45°/-45°/90°$$

Table 7.3 SCF at Circular Hole in Tension Panel (Laminate Data from Table 7.1)

Lay-Up		SCF
No. 0° Plies	No. ±45° Plies	Point A
24	0	6.6
16	8	4.1
12	12	3.5
8	16	3.0
0	24	2.0

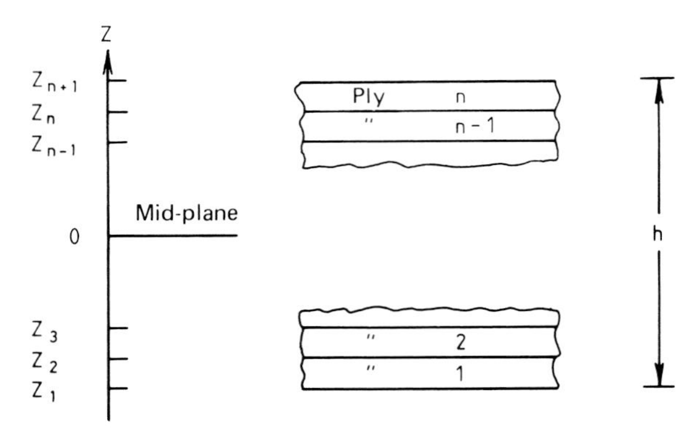

Fig. 7.9 Ply coordinates in thickness direction.

the order in the sequence being reversed at the midplane to preserve the symmetry. This would be written in code form as

$$[(0_2/\pm 45/90)_5]_s$$

Sometimes, in general discussions, a laminate is described by the percentages of its plies at various angles. Thus, the laminate of example 1 would be described as having "50% 0s, 50% ±45s." Similarly, the laminate of example 2 would be described as having "40% 0s, 40% ±45s, 20% 90s."

7.5 GENERAL LAMINATES SUBJECTED TO PLANE STRESS AND BENDING LOADS

Here the previous restriction to laminates that are symmetric about the midthickness plane will be dropped. It now becomes necessary to consider the plane stress and bending problems in conjunction, as in-plane loads can induce bending deformations and vice versa. Only the outline of the theory will be given below; for further details, including numerical examples, see Refs. 1–5 and 12.

General Theory

In distinction to the situation for symmetric laminates, the position of each ply in the laminate now is of importance. Thus, consider an n-ply laminate and denote by z the coordinate in the thickness direction, measured from the midthickness plane; the kth ply lies between z_k and z_{k+1} (Fig. 7.9). As before, the total thickness of the laminate will be denoted by h.

It is assumed that when a laminate is subjected to in-plane and/or bending loads, the strain at any point can be written in the form

$$\varepsilon_x = \varepsilon_x^* + K_x z \qquad \varepsilon_y = \varepsilon_y^* + K_y z \qquad \gamma_{xy} = \gamma_{xy}^* + K_{xy} z \tag{7.23}$$

where the starred quantities are the midplane strains and the K are the midplane curvatures (as in the bending of isotropic plates); both these sets of quantities are independent of z. Substituting from Eqs. (7.23) into Eqs. (7.7), it follows that the stresses in the kth ply will now be given by

$$\sigma_x(k) = Q_{11}(\theta_k)(\varepsilon_x^* + zK_x) + Q_{12}(\theta_k)(\varepsilon_y^* + zK_y) + Q_{16}(\theta_k)(\gamma_{xy}^* + zK_{xy})$$

$$\sigma_y(k) = Q_{12}(\theta_k)(\varepsilon_x^* + zK_x) + Q_{22}(\theta_k)(\varepsilon_y^* + zK_y) + Q_{26}(\theta_k)(\gamma_{xy}^* + zK_{xy})$$

$$\tau_{xy}(k) = Q_{16}(\theta_k)(\varepsilon_x^* + zK_x) + Q_{26}(\theta_k)(\varepsilon_y^* + zK_y) + Q_{66}(\theta_k)(\gamma_{xy}^* + zK_{xy})$$

$$(7.24)$$

Now introduce the stress resultants (in the form of forces per unit length and moments per unit length) defined by

$$N_x = \int \sigma_x \, \mathrm{d}z \qquad N_y = \int \sigma_y \, \mathrm{d}z \qquad N_{xy} = \int \tau_{xy} \, \mathrm{d}z$$

$$M_x = \int \sigma_x z \, \mathrm{d}z \qquad M_y = \int \sigma_y z \, \mathrm{d}z \qquad M_{xy} = \int \tau_{xy} z \, \mathrm{d}z \tag{7.25}$$

where all of the integrals are over the thickness of the laminate (i.e., from $z = -h/2$ to $z = h/2$); see Fig. 7.10. Since each of the integrals in Eqs. (7.25) can be written in forms such as

$$N_x = \sum_{k=1}^{n} \int_{z_k}^{z_{k+1}} \sigma_x(k) \, \mathrm{d}z, \qquad M_x = \sum_{k=1}^{n} \int_{z_k}^{z_{k+1}} \sigma_x(k) z \, \mathrm{d}z \tag{7.26}$$

it follows that substituting from Eqs. (7.24) into Eqs. (7.25), and performing some elementary integrations, leads to the result

$$\begin{bmatrix} A_{11} & A_{12} & A_{16} & B_{11} & B_{12} & B_{16} \\ A_{12} & A_{22} & A_{26} & B_{12} & B_{22} & B_{26} \\ A_{16} & A_{26} & A_{66} & B_{16} & B_{26} & B_{66} \\ B_{11} & B_{12} & B_{16} & D_{11} & D_{12} & D_{16} \\ B_{12} & B_{22} & B_{26} & D_{12} & D_{22} & D_{26} \\ B_{16} & B_{26} & B_{66} & D_{16} & D_{26} & D_{66} \end{bmatrix} \begin{bmatrix} \varepsilon_x^* \\ \varepsilon_y^* \\ \gamma_{xy}^* \\ K_x \\ K_y \\ K_{xy} \end{bmatrix} = \begin{bmatrix} N_x \\ N_y \\ N_{xy} \\ M_x \\ M_y \\ M_{xy} \end{bmatrix} \tag{7.27}$$

The elements in the above combined "extensional-bending stiffness matrix" are given by

$$A_{ij} = \sum_{k=1}^{n} Q_{ij}(\theta_k)(z_{k+1} - z_k) \tag{7.28}$$

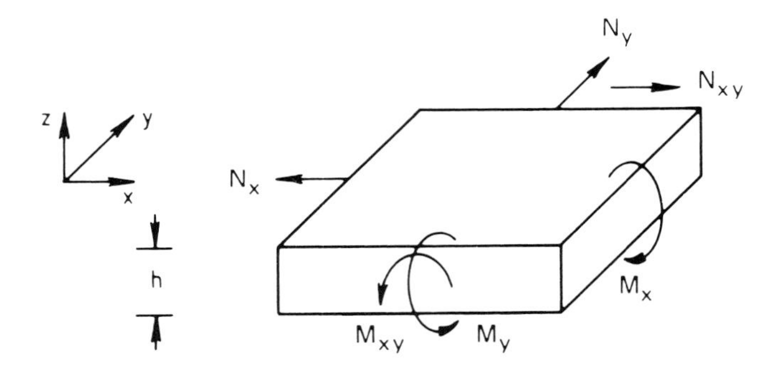

Fig. 7.10 Stress and moment resultants for a laminate.

$$B_{ij} = \sum_{k=1}^{n} Q_{ij}(\theta_k)(z_{k+1}^2 - z_k^2)/2 \tag{7.29}$$

$$D_{ij} = \sum_{k=1}^{n} Q_{ij}(\theta_k)(z_{k+1}^3 - z_k^3)/3 \tag{7.30}$$

[The above definition for A_{ij} is, of course, equivalent to that of Eq. (7.13).]

Equations (7.27) can be used to develop a theory for the stress analysis of general laminates, but this is quite formidable mathematically; it involves the solution of two simultaneous fourth order partial differential equations (see p. 3-43 of Ref. 7).

Uncoupling of Stiffness Matrix

It can be seen from Eqs. (7.27) that the plane stress and bending problems are coupled unless all the B_{ij} are zero. It follows from Eq. (7.29) that the B_{ij} are indeed zero for a symmetric lay-up. [For a symmetric pair of plies, if that ply below the midplane has coordinates $z_k = -a$ and $z_{k+1} = -b$, then its mate above the midplane will have $z_k = b$ and $z_{k+1} = a$; since each ply has the same Q_{ij}, there is a cancellation in the summation of Eq. (7.29).]

It is possible to achieve an approximate uncoupling of Eqs. (7.27) for a multiply unsymmetric laminate by making it in the form of a large number of repetitions of a given ply grouping. It can be seen intuitively that such a laminate will be symmetric in "the group of plies" if not in the individual plies; as long as the number of plies in the group is small compared with the total number of plies in the laminate, the B_{ij} will turn out to be small quantities. For example, a 48-ply laminate containing 24 groups of $\pm 45°$ plies laid up in the sequence $+/-/+/-/$, etc. (without there being symmetry about the midplane) would be expected to behave much as a symmetrical laminate of the same plies.

7.6 BENDING OF SYMMETRIC LAMINATES

The moment-curvature relation governing the bending of symmetric laminates out of their plane, as extracted from Eqs. (7.27), is

$$\begin{bmatrix} D_{11} & D_{12} & D_{16} \\ D_{12} & D_{22} & D_{26} \\ D_{16} & D_{26} & D_{66} \end{bmatrix} \begin{bmatrix} K_x \\ K_y \\ K_{xy} \end{bmatrix} = \begin{bmatrix} M_x \\ M_y \\ M_{xy} \end{bmatrix} \tag{7.31}$$

Analogously to the definition of orthotropy in plane stress, a laminate is said to be orthotropic in bending if $D_{16} = D_{26} = 0$. However, it is important to note that a laminate that is orthotropic in plane stress is not necessarily orthotropic in bending. For example, consider the four-ply laminate $[\pm 45]_s$. Here the coordinates for the $+45°$ plies may be written as $(-h/2, -h/4)$ and $(h/4, h/2)$, while those of the $-45°$ plies are $(-h/4, 0)$ and $(0, h/4)$; h, of course, is the laminate thickness. It is easy to establish that, while A_{16} and A_{26} are zero, D_{16} and D_{26} are not. On the other hand, a laminate containing only 0° and 90° plies will be orthotropic in both plane stress and bending. (It is also worth noting that for multiply laminates made of "groups of plies," as described in Sec. 5, if the group is orthotropic in plane stress, then the laminate will at least be approximately orthotropic in both plane stress and bending.)

For an orthotropic plate in bending, the deflection w satisfies the following equation:

$$D_{11}\frac{\partial^4 w}{\partial x^4} + 2(D_{12} + 2D_{66})\frac{\partial^4 w}{\partial x^2 \partial y^2} + D_{22}\frac{\partial^4 w}{\partial y^4} = q \tag{7.32}$$

where q is the applied pressure. Solutions of this equation can be found in Refs. 1 and 13; in both these references, the related problem of the buckling of laminated plates is also discussed.

7.7 FAILURE CRITERIA FOR LAMINATES

The problem considered here is broadly that of predicting the ultimate strength of a laminate under plane stress conditions, given relevant strength data for a single ply. This is quite a complex subject on which research is still actively proceeding and only the briefest of outlines can be given; however, detailed reviews can be found in Refs. 14 and 15. (At this stage, it is usual to rely largely on test data for laminate strength values.)

Using the theory of Sec. 4, the stresses in the individual plies of a laminate can be determined. Further, using the inverse transformation to Eq. (7.5), the stresses at any point in a ply can be referred to the material axes. Hence, the problem can be reduced to establishing a criterion for the ultimate strength of a single ply with the stresses referred to the material axes. As before, these stresses will be denoted by σ_1, σ_2, and τ_{12}.

The available data (established mainly from unidirectional laminate tests) will usually be as follows:

F_{1T}, F_{1C} = ultimate strength in tension and compression, respectively, in fiber direction

F_{2T}, F_{2C} = ultimate strength in tension and compression, respectively, in transverse direction

F_{12} = ultimate shear strength

Maximum Stress Criterion

The simplest criterion is to assume that failure occurs only when any of the following conditions is violated:

$$-F_{1C} \leq \sigma_1 \leq F_{1T} \qquad -F_{2C} \leq \sigma_2 \leq F_{2T} \qquad -F_{12} \leq \tau_{12} \leq F_{12} \tag{7.33}$$

Since this criterion does not allow for any interaction between the stress components, it is, in general, nonconservative.

Maximum Strain Criterion

It is possible to establish by test ultimate strains for a single ply (rather than ultimate stresses) and then set up a strain criterion quite analogous to Eq. (7.33). However, this also is, in general, nonconservative.

Tsai-Hill/Hoffman Criterion

The criterion that is probably most commonly used is one developed by Tsai from a formula proposed by Hill in a different connection (namely, the yielding of an anisotropic metal). In its original form, the Tsai-Hill criterion did not allow explicitly for the differences between the tensile and compressive ultimate strengths; Hoffman extended the criterion to cover that situation (which is the usual one). According to the Tsai-Hill/Hoffman criterion, failure occurs if the following inequality is violated:

$$\sigma_1^2/(F_{1T}F_{1C}) - \sigma_1\sigma_2/(F_{1T}F_{1C}) + \sigma_2^2/(F_{2T}F_{2C})$$

$$+ (1/F_{1T} - 1/F_{1C})\sigma_1 + (1/F_{2T} - 1/F_{2C})\sigma_2 + \tau_{12}^2/F_{12}^2 \leq 1 \tag{7.34}$$

This criterion does allow for stress interaction effects, as can be seen from Fig. 7.11, where the criteria [Eqs. (7.33) and (7.34)] are compared for the following simplified case:

$$F_{1T} = F_{1C} = 1400 \text{ MPa}$$

$$F_{2T} = F_{2C} = 140 \text{ MPa}$$

$$\tau_{12} = 0$$

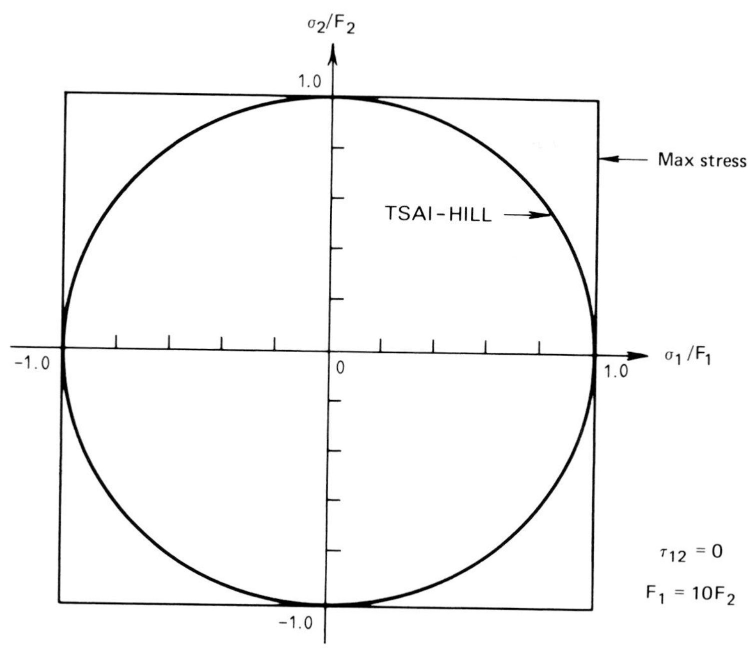

Fig. 7.11 Comparison of maximum stress and Tsai-Hill failure criteria for a single ply.

There are some theoretical objections to the criterion of Eq. (7.34) on the grounds of a lack of "tensor invariance"; see Refs. 2 and 15. A more refined criterion, often known as the Tsai-Wu criterion, which does not suffer that objection has been developed in Ref. 15, but it contains another material constant that is difficult to determine accurately.

References

[1]Jones, R. M., *Mechanics of Composite Materials*, McGraw-Hill Kogakusha Ltd., Tokyo, 1975.

[2]Tsai, S. W. and Hahn, H. T., *Introduction to Composite Materials*, Technomic Publishing Co., Westport, CT, 1980.

[3]Calcote, L. R., *The Analysis of Laminated Composite Structures*, Van Nostrand Reinhold, New York, 1969.

[4]Ashton, J. E., Halpin, J. C., and Petit, P. H., *Primer on Composite Materials: Analysis*, Technomic Publishing Co., Westport, CT, 1969.

[5]Agarwal, B. D. and Broutman, L. J., *Analysis and Performance of Fiber Composites*, John Wiley & Sons, New York, 1980.

[6]Broutman, L. J. and Krock, R. H. (eds.), *Composite Materials*, Vols. 7 and 8 ("Structural Design and Analysis," Pts. I and II edited by C. C. Chamis), Academic Press, New York, 1975.

[7]*Plastics for Aerospace Vehicles*, Pt. 1, "Reinforced Plastics," MIL-HDBK-17A, U.S. Department of Defense, Washington, DC, 1971.

[8]Hearmon, R. F. S. *An Introduction to Applied Anisotropic Elasticity*, Oxford University Press, London, 1961.

[9]Lekhnitskii, S. G., *Theory of Elasticity of an Anisotropic Elastic Body*, Holden-Day, San Francisco, 1963.

[10]Lekhnitskii, S. G., *Anisotropic Plates*, Gordon and Breach, New York, 1968.

[11]Tsai, S. W., "Mechanics of Composite Materials," Pts. I and II, AFML-TR-66-149, 1966.

[12]Hoskin, B. C. and Green, B. I., "Stress Distributions in Fiber Reinforced Plastic Laminates under Simple Loadings," ARL Note SM 396, 1973.

[13]Vinson, J. R. and Chou, T-W, *Composite Materials and Their Use in Structures*, Applied Science Publishers, London, 1975.

[14]Vicaris, A. A. and Toland, R. H., "Failure Criteria and Failure Analysis of Composite Structural Components," *Composite Materials*, Vol. 7, pp. 51–97. (See Ref. 6.)

[15]Wu, E. M., "Phenomenological Anisotropic Failure Criterion," *Composite Materials*, Vol. 2, pp. 353–431. (See Ref. 6.)

8. JOINING ADVANCED FIBER COMPOSITES

8.1 INTRODUCTION

Most structures consist of an assembly of a number of individual elements connected to form a load transmission path. These connections or joints are potentially the weakest points in the structure and may determine its viability. This chapter deals with the joining of advanced fiber composites, particularly graphite/epoxy, for aircraft applications where the joints are not subjected to significant bending moments.

In general, it is desirable to minimize the number of joints in a structure in order to minimize both its weight and cost. Fiber composites have an important advantage over metals in this respect, since large one-piece components are readily produced. Nevertheless, joints will be required to transmit loads in and out of the composite structure. Usually, joints employing metallic members, either aluminum or titanium alloys, are used for this purpose. Metallic members are particularly employed where significant concentrated loads occur or where significant through-the-thickness stresses arise.

The types of composite-to-metal joints can be classified as: (1) adhesively bonded; (2) mechanical, either fastened (bolted, riveted, etc.) or wedge loaded; and (3) combinations of types 1 and 2. In the first type, loads are carried by the surface of the joint elements in shear through a layer of adhesive or resin. In the second type, the loads are carried by the joint elements in compression on the contacting faces.

A variety of aspects must be considered when comparing the two major procedures, bolting and adhesive bonding, for joining advanced fiber composites to metals. Some of the salient points of comparison are listed in Table 8.1 which shows that adhesive bonding, where appropriate, produces the most structurally efficient and aerodynamically smooth joint. In general, however, bonding is limited to joining material of equivalent thickness and strength to about 4 mm of a high-strength aluminum alloy, unless joints of complex configuration are employed.

8.2 ADHESIVE BONDING

General Design Considerations

In adhesive bonding, the aim is to transfer the load smoothly from one adherend to the other, minimizing the peak shear stresses and peel stresses in the adhesive layer. Figure 8.1 illustrates the elemental joint configurations

Table 8.1 Comparison of Bonded and Bolted Composite-to-Metal Joints for Aircraft Applications

Advantages	Disadvantages
	Bonded joints
No stress concentration in adherends	Limits to thickness that can be jointed with simple joint configuration
Stiff connection	Inspection difficult.
Excellent fatigue properties	Prone to environmental degradation
No fretting problems	Requires high level of process control
Sealed against corrosion	Sensitive to peel and through-thickness stresses
Smooth surface contour	Residual stress problems when joining dissimilar materials
Relatively light weight	Cannot be disassembled
Damage tolerant	
	Bolted joints
Positive connection	Considerable stress concentration
No thickness limitations	Relatively compliant connection
Simple process	Relatively poor fatigue properties
Simple inspection procedure	Hole formation may cause damage to composite
Simple joint configuration	Prone to fretting
Not environmentally sensitive	Prone to corrosion
Provides through-thickness reinforcement—not sensitive to peel stress	
No residual stress problems	
Can be disassembled	

appropriate for joining advanced fiber composites to metals. To minimize manufacturing costs, the simplest joint configuration is chosen that will develop the required strength. For thin section joints, the choice is usually the lap or double-lap joint, both of which require minimal machining and involve simple manufacturing procedures.

Four main modes of failure of a joint must be considered for design purposes: shear of the adhesive system, peel of the adhesive system, peel of the composite, and tension or compression failure of the adherends. Most joint design procedures aim for tension or compression failure of the adherends as the lowest static failure loads, since this ensures a joint efficiency of at least 100% and that the bonded region does not become the weak link. This approach also usually ensures that the adhesive is not the weak link under cyclic loading.

The design parameters usually include (1) stiffness and strength of the adherends; (2) yield strength, stiffness, and strain to failure of the adhesive under shear loading, allowing for temperature and degradation due to representative service environment; (3) mismatch in stiffness between the adherends; (4) mismatch in thermal expansion coefficient between the adherends; and (5) the direction of the stresses.

Most of the factors influencing the behavior of structural adhesive joints are reviewed in Ref. 1. The stress-strain behavior of the adhesive is a most important materials property affecting joint strength. This information may be obtained from tensile tests on thick adherend lap joint specimens or from "napkin-ring" tests (end-bonded metal cylinders, tested in torsion). Figure 8.2a shows curves for a typical structural film adhesive obtained from the thick adherend lap test and illustrates the degree of plasticity exhibited by these materials; these results were obtained after moisture conditioning the adhesive to representative environmental conditions. Similar tests may be performed to determine effects of cyclic or prolonged static loading. The durability of the joint in service is most reliably assessed by subjecting structural detail specimens to realistic loading and environmental condi-

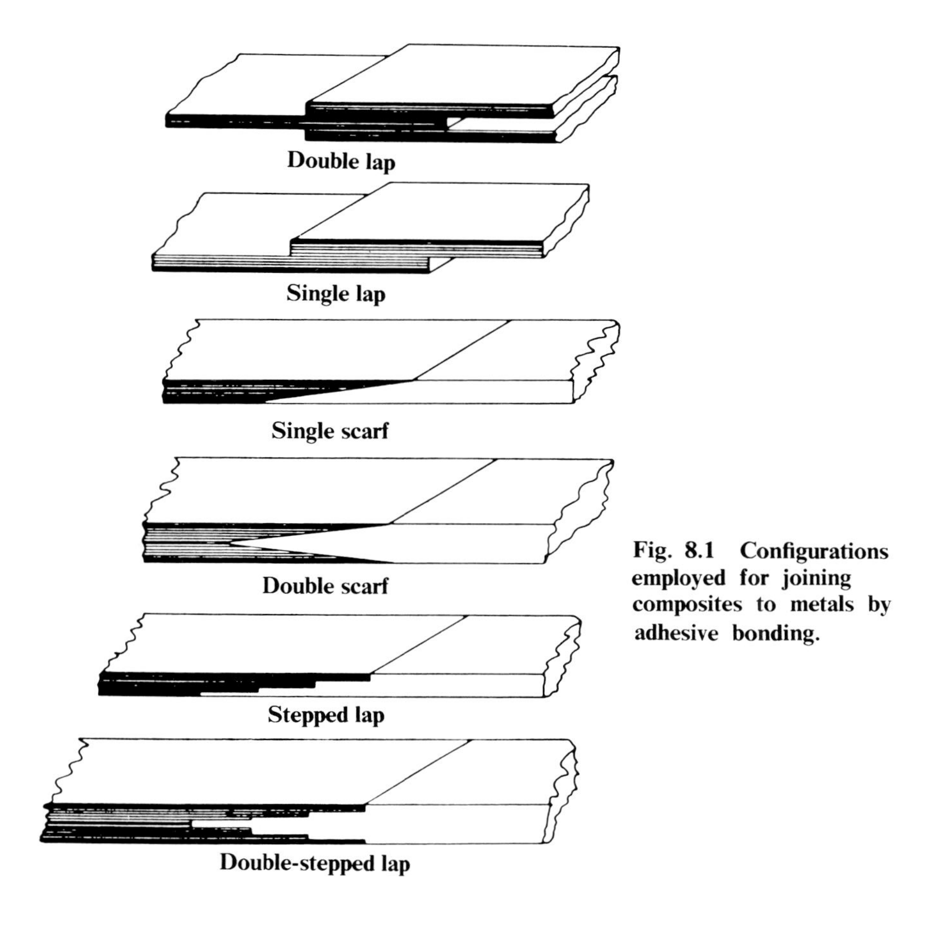

Fig. 8.1 Configurations employed for joining composites to metals by adhesive bonding.

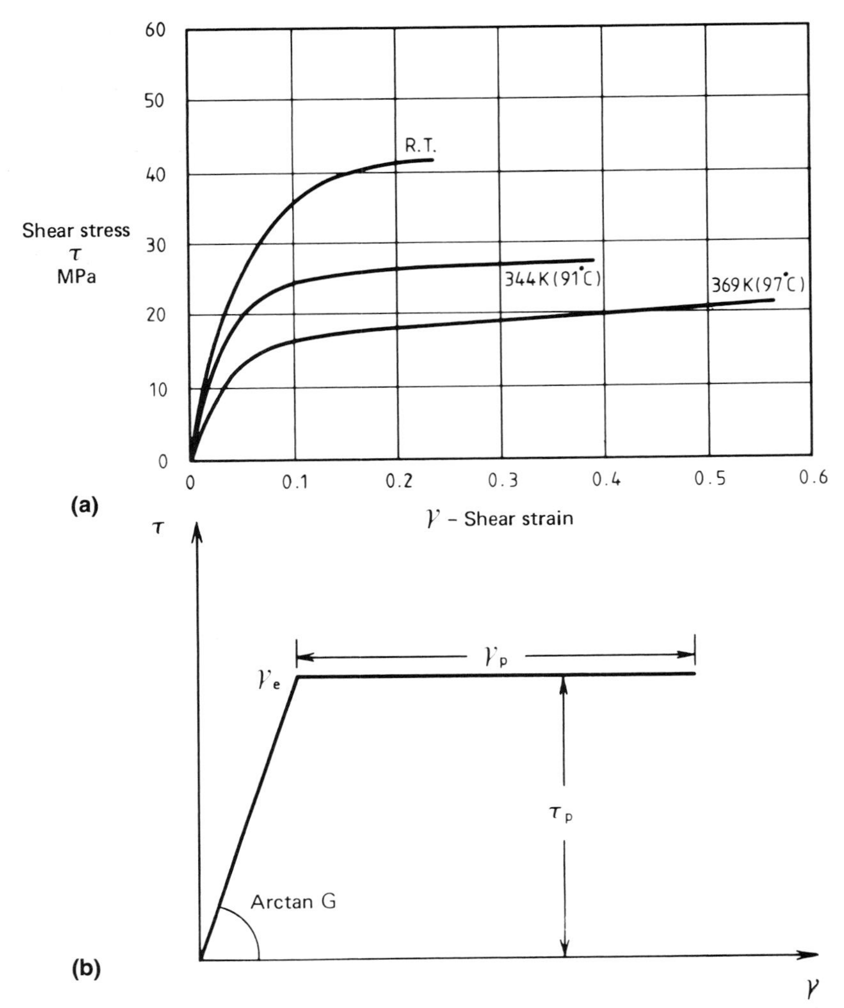

Fig. 8.2 Shear stress/shear strain curves: (a) plots of shear stress vs shear strain at various temperatures for adhesive FM 300 after exposure to moisture (results obtained from a thick adherend test); (b) idealization of a typical shear stress/shear strain curve for analytical purposes: the aim is to have an area under curve $\tau_p(\frac{1}{2}\gamma_e + \gamma_p)$ similar to that under the real curve.

tions. Simple coupon tests are best employed to aid initial design; however, even with sophisticated analytical procedures, these tests should not be relied on to predict allowable failure loads in composite-to-metal joints.

The most comprehensive design studies on adhesive joints, particularly for advanced fiber composite-to-metal joints have been performed by Hart-Smith.[2,3] These analytical studies are based on previous classical work (e.g., by Volkersen, and Reissner and Goland), but extended to take account of yielding in the adhesive,* as shown in Fig. 8.2a, using the idealized curves shown in Fig. 8.2b. Based on these studies, Fig. 8.3 schematically illustrates the strength limitations of various joint designs as a function of adherend stiffness. This indicates that single-lap joints (if unsupported) are limited to very thin adherends and that double-lap joints can be greatly increased in thickness capability by tapering the ends of the outer adherends to avoid peel failures. However, for very thick adherends, scarf or stepped-lap joint configurations must be employed.

In the following sections, some of the factors influencing the double-lap joint are first considered in some detail—this is the simplest effective joint illustrating most of the important features of composite-to-metal joints. This is followed by a brief description of the other types of bonded joints.

Double-Lap Joint

The model and notation used in Ref. 2 to estimate the shear stress distribution and load capacity of a double-lap joint are shown in Fig. 8.4; idealized stress-strain behavior of the adhesive is assumed (Fig. 8.2b). The following discussion initially considers the joint overlap length, then briefly considers each of the following parameters in terms of their effect on load capacity of the joint: adhesive plasticity, joint stiffness, adherend stiffness imbalance, and adherend thermal expansion mismatch. (For the detailed derivation of the formulas cited below, see Refs. 2 and 3.)

Overlap length. The idealized shear stress distribution in a well-designed stiffness-balanced joint (i.e., one where $E_i t_i = 2E_0 t_0$) is illustrated schematically in Fig. 8.5. This shows a fully developed elastic trough of length $6/\lambda$ (where λ is defined in Fig. 8.5 and governs the exponential decay of the elastic shear stress) and having two zones at the ends of the joint in which the adhesive is stressed to its yield stress τ_p. The load-carrying capacity P of the joint is obtained from the area under the shear stress curve. The maximum load (per unit width) that can be carried by the adhesive elastically is τ_p/λ; the rest of the load can be carried by the adhesive plastically, provided that the plastic strain in shear does not exceed γ_p. The maximum joint strength occurs when the plastic zone and the elastic zone are fully developed; increases in overlap length above this do not increase the joint's load-carrying capacity.

While, in most cases, the applied load could be carried with a shorter overlap length, there are several reasons why a length greater than the maximum length indicated in Fig. 8.5 should be employed. A fully developed elastic zone serves (1) to stabilize the joint against creep in the adhesive by maintaining very low shear stresses in the central region; (2) to provide allowance for manufacturing defects such as voids and debonds; and (3) to allow for service deterioration such as fatigue damage, debond-

*The main weakness in this theory is that it does not account for viscoelastic effects in the adhesive that may have an important bearing on situations involving sustained loading.

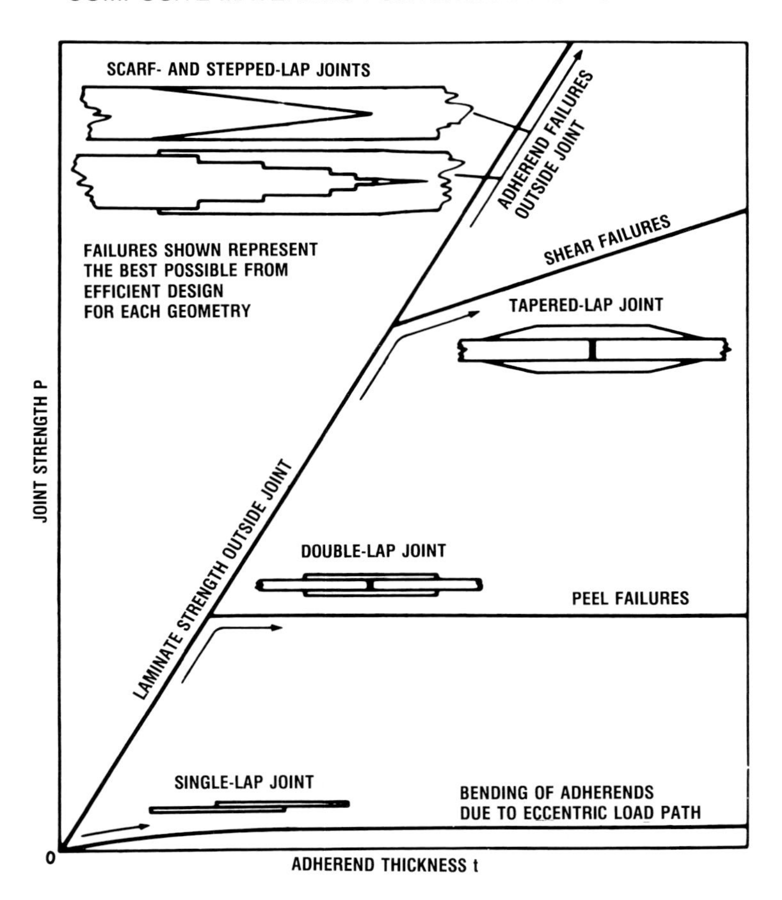

Fig. 8.3 Schematic of strength limitation of various joint designs as a function of adherend stiffness. (Taken from Ref. 2.)

ing, and general environmental degradation. In effect, a suitably long overlap length serves to make the joint damage tolerant.

For the balanced joint of Fig. 8.5 with $E_0 = E_i = E$ and $t_0 = t_i/2 = t$, the required minimum overlap length for loading up to the ultimate strength σ_{ult} of the adherends (i.e., to obtain 100% joint efficiency) is given by

$$l_{min} = t\sigma_{ult}/\tau_p + 2/\lambda \tag{8.1}$$

where t is the adherend thickness. The adhesive properties for the hot/wet condition should be assumed for this calculation; an allowance of at least 25% is usually made for uncertainties. (For adherends with stiffness and strength similar to 2024T3 aluminum alloy and thickness t of 1.5 mm, joined with a modified epoxy structural adhesive, l_{min} is of the order of 40 mm.)

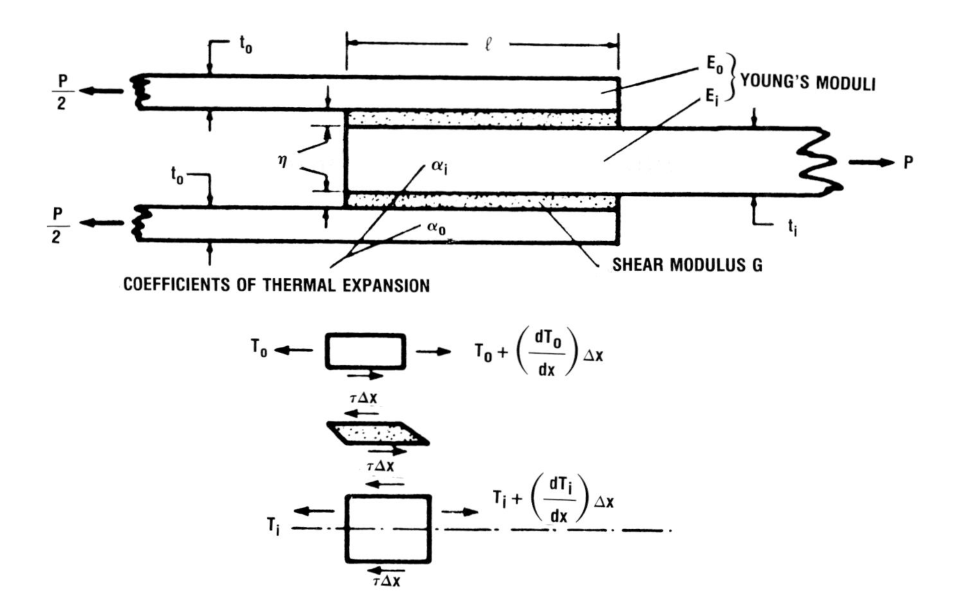

Fig. 8.4 Part of model employed in Ref. 2 to analyze the double-lap joint.

Effect of adhesive plasticity on strength. The maximum load-carrying capacity P per unit width of a joint in shear in which the adherends have equal stiffness and expansion coefficient is given by, using the notation of Fig. 8.5,

$$P = 4\left[\eta\tau_p\left(\tfrac{1}{2}\gamma_e + \gamma_p\right)Et\right]^{\frac{1}{2}} \tag{8.2}$$

The quantity $\tau_p(\frac{1}{2}\gamma_e + \gamma_p)$, Fig. 8.2b is the area under the stress-strain curve and is the work to failure per unit volume of the adhesive. The load-carrying capacity increases as the square root of this term, which is dominated by $\tau_p\gamma_p$ for a ductile adhesive. Typically, γ_p/γ_e is 20–50 for an epoxy-nitrile adhesive at ambient temperature. Further, since the quantity $\tau_p(\frac{1}{2}\gamma_e + \gamma_p)$ does not vary much with temperature (see Fig. 8.2a), joint strength does not vary much with temperature.

Effect of joint stiffness on maximum strength. Equation 8.2 shows that, for a stiffness balanced joint, P is proportional to $\sqrt{(Et)}$. Thus, the load-carrying capacity of the joint increases as a function of the square root of adherend stiffness. Since, however, the adherend strength increases proportionally to t, the load-bearing capabilities of the adherends increase more rapidly than the joint strength. In practice, assuming adherends with the strength and stiffness of aluminum alloy 2024T3 and ignoring the peel effects (to be discussed later), the cross-over point occurs at a thickness of

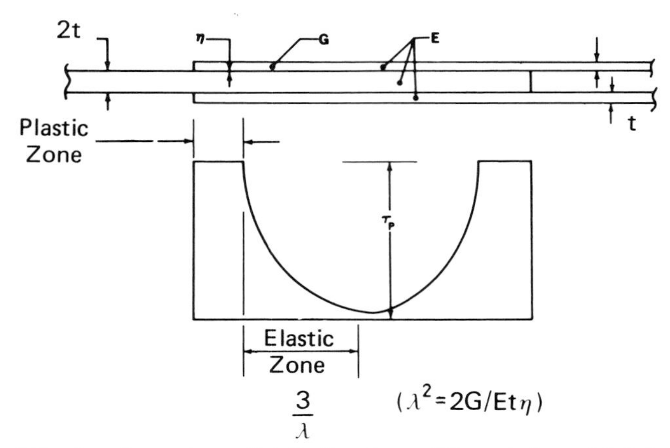

Fig. 8.5 Adhesive shear stress distribution in a stiffness-balanced joint of optimum length.

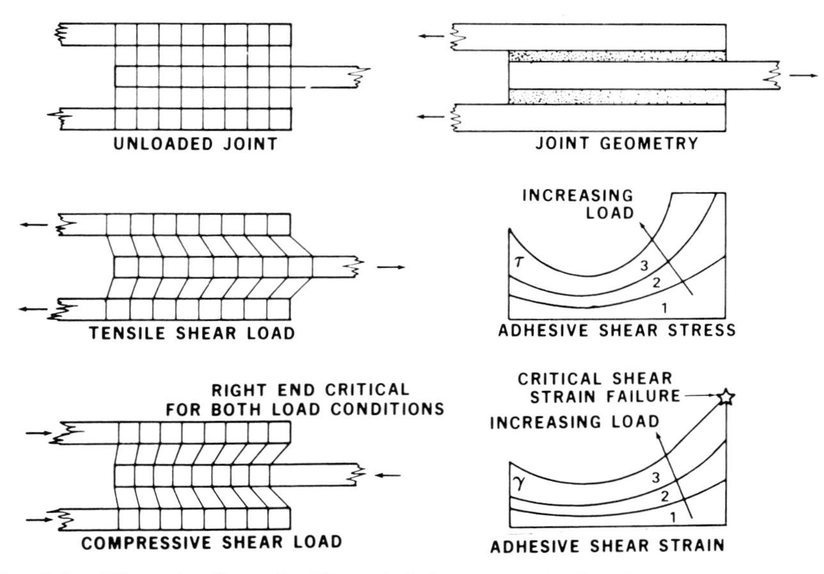

Fig. 8.6 Effect of adherend stiffness imbalance on adhesive shear strength. (Taken from Ref. 2.)

about 6 mm where the load capacity of the joint is about 2.5 MN/m (15,000 lb/in.) and the strength of the adherends is about 2.9 MN/m.

Effect of adhesive thickness. Equation (8.2) shows that the maximum load-carrying capacity P increases in proportion to $\eta^{\frac{1}{2}}$, where η is the thickness of the adhesive layer. For most practical joints, the adhesive layer thickness is held to 0.13–0.26 mm. Thicker layers tend to have a high void content and so are not a practical means of increasing strength. However, excessively thin layers reduce joint strength and must be avoided during manufacture: the presence of carrier fibers in structural film adhesives is highly beneficial in preventing excessive adhesive extrusion and, therefore, the formation of thin layers during joint manufacture.

Effect of stiffness imbalance of adherends. Figure 8.6 schematically shows the situation that arises when the outer members of a double-lap joint are stiffer than the inner member. The highest stress occurs at the end of the joint from which the more compliant member projects, irrespective of the direction of loading. This results in a reduction of the strength of the joint. Let S be the ratio of the stiffness of the inner to the two outer adherends; then $S = 1$ is a balanced joint. Thus, the strength reduction for $S > 1$ is approximately $\sqrt{(1 + 1/S)/2}$ and, for $S < 1$, the reduction factor is $\sqrt{S(1 + S)/2}$.

Influence of thermal expansion mismatch. This is an inherent problem with composite-to-metal joints since, for aluminum alloys, the expansion coefficient is $\alpha = 23 \times 10^{-6} {}^\circ C^{-1}$, while for graphite/epoxy, typically, $\alpha = 0.5 \times 10^{-6} {}^\circ C^{-1}$. For a stiffness-balanced joint experiencing a temperature change (service temperature minus adhesive cure temperature) of ΔT, the reduction in the load capacity of the joint is approximately $2E_0 t_0 \, \Delta\alpha \, \Delta T$.

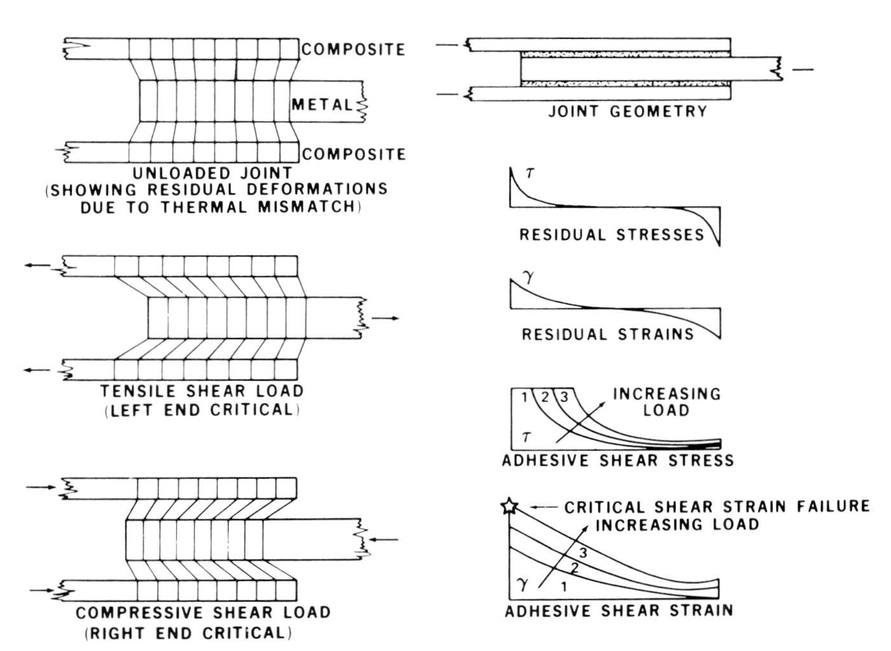

Fig. 8.7 Effects of adherend thermal expansion mismatch on adhesive shear strength. The lower two diagrams on the right refer to tensile loading. (Taken from Ref. 2.)

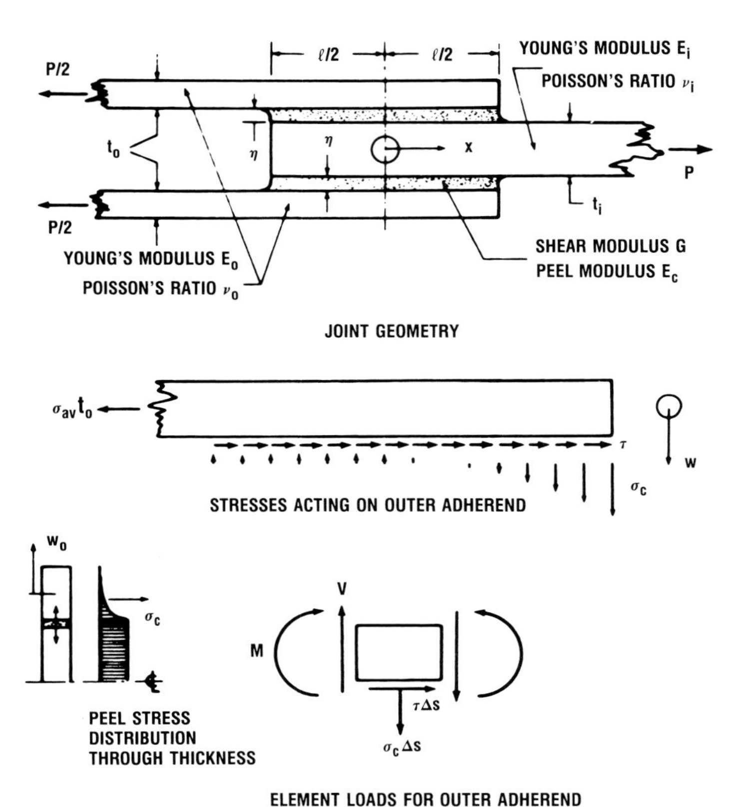

Fig. 8.8 Model used in Ref. 2 to analyze for peel stresses.

However, the critical end of the joint depends on the direction of loading as illustrated schematically in Fig. 8.7.

Effect of combined thermal expansion mismatch and stiffness imbalance. For this general case, and referring again to the notation of Fig. 8.4, the critical end of the joint cannot be obtained by inspection, because the strength is given by the lesser of two values,

$$P = (\alpha_0 - \alpha_i)\,\Delta T E_i t_i + \left[2\eta\tau_p\left(\tfrac{1}{2}\gamma_e + \gamma_p\right)2E_i t_i\left(1 + \frac{E_i t_i}{2E_0 t_0}\right)\right]^{\frac{1}{2}} \quad (8.3)$$

or

$$P = (\alpha_i - \alpha_0)\,\Delta T\, 2E_0 t_0 + \left[2\eta\tau_p\left(\tfrac{1}{2}\gamma_e + \gamma_p\right)4E_0 t_0\left(1 + \frac{2E_0 t_0}{E_i t_i}\right)\right]^{\frac{1}{2}} \quad (8.4)$$

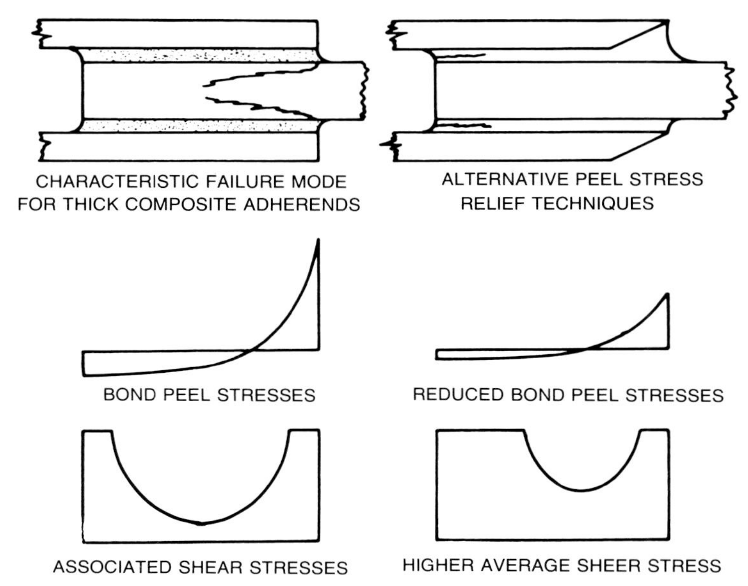

Fig. 8.9 Peeling failure modes in a composite inner adherend and geometric methods of reducing peel stresses. (Taken from Ref. 2.)

Peel stress limitation on strength and methods of alleviation. The mathematical model employed in Ref. 2 to estimate peel stress is shown in Fig. 8.8. The main simplifying assumption made in the analysis is that the shear stress τ at the ends of the adherends is constant. This is appropriate to a situation in which the adhesive has yielded. Peel stresses under tension loading arise at the outer edges of the outer adherends as a result of the tendency of the outer adherends to bend away from the inner adherend under the moment produced by the shear stresses. The bending is a result of the development of a peeling stress σ_c in the adhesive layer and adherends that is approximately given by

$$\sigma_c = \tau \left(\frac{3E_c' t_0}{E_0 \eta} \right)^{\frac{1}{4}} \tag{8.5}$$

where E_c' is the *effective* transverse stiffness of the adhesive *system*. Thus, for a given adhesive system, σ_c is increased with the increasing thickness of the outer adherend and is reduced with increasing modulus of the outer adherend and increasing thickness of the adhesive layer. Use of a composite in place of a metal for the outer adherend reduces E_c' and, thus, the peel stress. Peel stresses can be alleviated by control of these parameters and by geometrical means such as scarfing to locally reduce t_0, as illustrated in Fig. 8.9. The influence of scarfing in reducing the strength cutoff due to peel is shown schematically in Fig. 8.3.

The peel problem is particularly serious with fiber composites since they generally have a low transverse strength, even compared with the adhesive, particularly if the adhesive is of the ductile peel-resistant type.

Design of Other Bonded Joints

Single-lap joints. The behavior of single-lap joints is dominated by peeling effects arising from the bending moments produced by the eccentricity in the load path. Thus, these joints should generally not be employed for fiber composite adherends since they usually have poor transverse strength.

These problems do not arise, however, in a single-lap joint supported against bending, e.g., by a substructure to which the joint may be attached or fastened. In this case, the double-lap joint analysis is applicable, considering the single-lap joint as a symmetrical half of the double-lap joint.

Scarf joints. The shear stress in the adhesive layer is reasonably uniform in a scarf joint, provided that the adherends have equal stiffness and thermal expansion coefficients and that the scarf is taken to a sharp edge. Peel stresses and transverse stresses σ_T are also very low at low scarf angles θ.

Simple theory gives for the shear stress along the bond line

$$\tau = P \sin 2\theta / 2t \tag{8.6a}$$

and for the stress normal to the bond line

$$\sigma_T = P \sin^2\theta / t \tag{8.6b}$$

For small scarf angles, the conditions for failure in the adherends are given by

$$\theta < \tau_p / \sigma_{\text{ult}} \, \text{rad} \tag{8.7}$$

where, as before, σ_{ult} is the ultimate stress for the adherends. However, if the adherends are dissimilar, as in a composite-to-metal joint, the shear stress in the adhesive is not uniform and design becomes more complex. Where joints have adherends with different stiffness, the highest stress occurs at the end from which the more compliant adherend extends. For either elastic or plastic analysis, the ratio of the average shear stress to the peak shear stress asymptotes to the adherend stiffness ratio. Problems with stiffness imbalance and thermal expansion mismatch are greatly reduced by employing a multiple scarf configuration. As shown schematically in Fig. 8.3, scarf joints are capable of joining adherends of any thickness.

Stepped-lap joints. Stepped-lap joints can be analyzed as a succession of double laps. In common with the double-lap joint, the stepped-lap joint has a nonuniform shear stress distribution with high stresses at the ends of each step. However, with suitable design, the stepped-lap joint is capable of joining adherends of considerable thickness. In order to increase the load-carrying capacity of a stepped-lap joint, it is not usually sufficient just to increase the length of the steps, since the load-carrying capacity of each step does not increase indefinitely, as discussed earlier for double-lap joints; instead, it is necessary to increase the number of steps. Indeed, to avoid overloading some of the end steps of the inner adherend, it may be necessary to reduce their length. The results of an elastic analysis of a stepped-lap joint (such as is used for the wing-to-fuselage joint on the F-18) is illustrated in Fig. 8.10. Peel stresses are not usually a problem because of the low thickness at the ends of the outer adherend.

Bonded and bolted joints. In general, an adhesive joint is much stiffer than a bolted (or riveted) joint. Thus, it is not possible to design a bonded and bolted joint where the load is effectively shared between the bonded and fastened regions. Fasteners may be used effectively, however, in thick section joints (where failure would normally occur preferentially in the adhesive) to contain local adhesive failure where design stresses are exceeded for any reason (e.g., due to overload or local bond flaws). For this

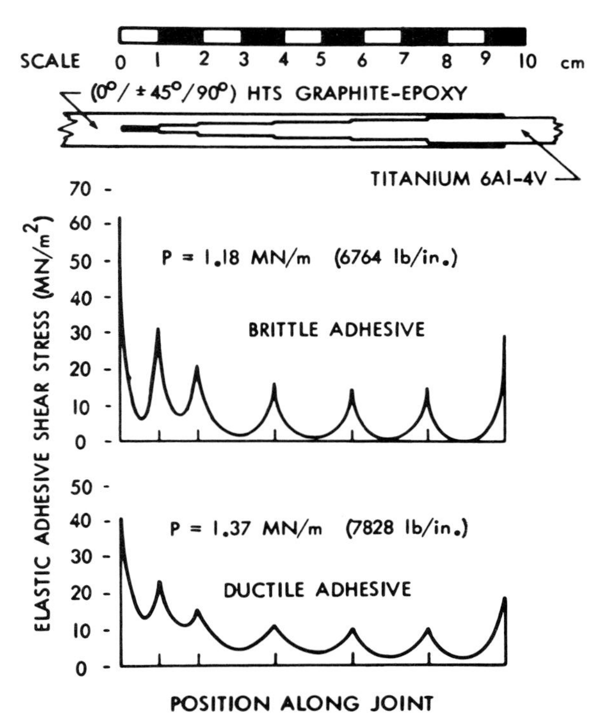

Fig. 8.10 Stepped-lap joint employed to join titanium to graphite/epoxy and the analytical shear stress distribution in the adhesive for two adhesive types. (Taken from Ref. 2.)

purpose, the fasteners should be situated away from high-stress regions in the adhesives—the center of the elastic trough in a lap joint is an ideal region. An alternative view, leading to similar conclusions for placement of the fasteners, is that use of adhesive bonding alleviates load in the fastened region, reducing the danger of fatigue failure from the fastener holes.

Rivets are sometimes employed at the ends of a lap joint to reduce peeling stresses. Although mechanically this approach may be effective, particularly for single-lap joints, it is a dangerous procedure in practice since the presence of fastener holes allows entry of moisture and other fluids into the critical high-shear-stress region and can result in environmentally induced bond failure. It is much safer to use the other procedures for reducing peel stresses, referred to earlier.

Materials Aspects

Optimum ply configuration. Generally for a bonded joint, the ply content (e.g., proportion of 0°, ±45°, and 90° plies) will be determined by factors other than joint design. However, the strength of the bonded joint (all other aspects taken as constant) depends on the orientation of the fibers at the joint interface. The shear strength of the interface is maximum when the outer layer of fibers is oriented in the same direction as the applied load and is minimum when the fibers are oriented perpendicular to the applied load. Thus, for a given ply content, the ply configuration should provide 0° fibers on the outer surface for lap joints and on the steps of a stepped-lap joint. No particular ply configuration is necessarily optimum for a scarf joint.

Adhesives. Structural film adhesives are usually employed for forming bonded joints between advanced fiber composite and metal adherends. These adhesives consist of a precatalyzed modified epoxy such as an epoxy-nitrile or, for higher-temperature applications, an epoxy-phenolic supported by a glass or polymer fiber, usually polyester or nylon, in mat or woven form. Cure temperatures vary from about 120°C for epoxy-nitrile to 180°C for epoxy-phenolic. The epoxy-nitrile adhesives are formulated to provide high peel resistance. This results from a dispersion of chemically bonded nitrile-rubber particles in the matrix, which consists of a dilute solid solution of the rubber in epoxy resin. These ductile adhesives suffer a marked loss in shear strength and stiffness above about 80°C. The epoxy-phenolic types of adhesive are suitable for use at temperatures up to about 120°C, but are brittle and therefore have relatively poor peel resistance at lower temperatures.

The carrier fibers in structural adhesive films serve several purposes, including providing mechanical strength to the uncured resin for handling purposes, controlling flow and thus inhibiting overthinning of the resin during joint formation, improving peel strength, and insulating the adherends in the event of complete resin squeeze-out; this is important where graphite/epoxy is joined to aluminum alloys, since electrical contact can result in severe galvanic corrosion of the aluminum. However, the carrier fibers can encourage wicking of moisture into the adhesive layer and result in degradation of the metal/adhesive bond. Woven fibers produce more of a problem than mat in this respect, since they provide a continuous diffusion path.

Surface treatment. The most critical step in joint formation is the production of the correct type of surface for bonding. To ensure good bond strength and durability, metals such as aluminum and titanium alloys are generally degreased, etched, and anodized, to form a stable oxide film. Prior to bonding, the surfaces are coated with a thin layer of a corrosion inhibiting primer—usually a low solubility chromate in an epoxy resin.

Fortunately, in the case of a cured epoxy matrix composite, production of a roughened resin surface, clean from surface contaminants (particularly release films) is all that is usually required. This can be achieved by degreasing and abrading the area with abrasive paper or by blasting with a grit such as aluminum oxide to remove the contaminated surface and increase surface energy. Alternatively, during manufacture, a layer of woven nylon cloth is incorporated into the surface of the composite. Prior to bonding, the nylon ply is peeled from the composite, exposing a clean surface ready for bonding. The nature of the resulting surface can be controlled to some extent by the weave of the nylon peel ply. However, the resulting surface is usually very rough and tends to entrap air bubbles. There is also a serious danger that small amounts of the peel ply may be left on the surface. It is generally considered that abrasion produces a more reliable surface for bonding.

The requirement for producing a clean surface in the composite (with all the inherent dangers) can be avoided by cocuring the composite with the adhesive. This involves consolidation of the composite at the same time as the adhesive is cured and effectively removes one adhesive/adherend interface.

Joint Manufacture

Precured composite. After surface treatment of composite and metal surfaces, film adhesive is placed between the faying surfaces and the parts are then supported on a special tool. Thermocouples are placed in position to monitor the temperature and temperature distribution, as well as the heat and pressure applied either in a heated press for relatively small simple components or in an autoclave for large or complex components. When autoclave curing is used, a vacuum bag is formed over the tool surface to hold the parts in place prior to the cure and to remove air and volatiles. The vacuum is released before gelling of the adhesive and the application of full pressure.

Cocuring. Here, the first half of a composite component is laid up with prepreg onto a tool with the configuration required to form the end shape, e.g., steps for a stepped-lap joint. In order to minimize movement during the joint formation, the composite lay-up, if consisting of many plies (say 20 or

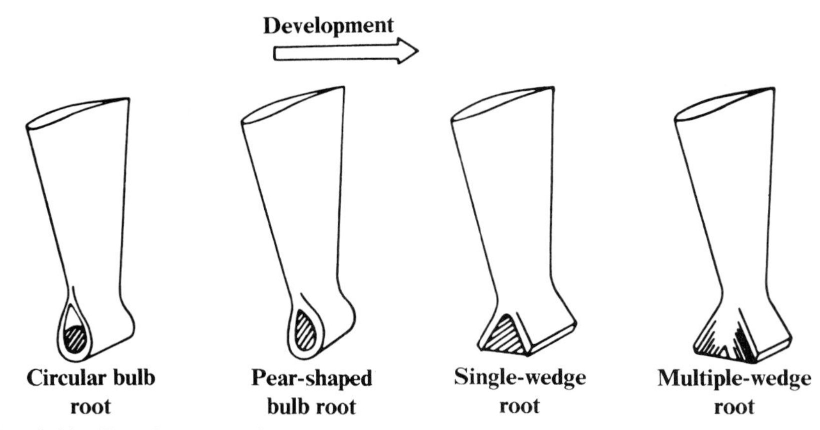

Fig. 8.11 Development of root configurations for fan blades, based on a wedging action.

so) is usually debulked at several stages by vacuum bagging operations. The contoured metal part, after appropriate surface treatment and coating with adhesive, is placed in position. Finally, the second half of the composite component is formed from prepreg laid up over the metal and the composite inner layer. A further vacuum bag is then placed over the part and the assembly is cured under normal autoclave conditions.

For this procedure to be successful, the adhesive and resin must be chemically compatible and must cure under a similar temperature, pressure, and time sequence.

Quality Control

Surface treatment. No simple direct methods are available for assessing the quality of surfaces after surface treatment. Thus, close analytical control is maintained over all chemical processes and control specimens are periodically surface treated, bonded, and tested. In addition, test coupons are usually manufactured along with bonded components (often cut from the component panel) and evaluated using durability and other tests.

Recently, the Fokker Aircraft Company produced an instrument it claims measures the quality of surface-treated metallic components by measuring the surface free energy. Polarized light tests can be used in some cases to establish the formation of satisfactory oxide films during anodizing of metal surfaces. Other surface examination techniques, such as ellipsometry, can be used to detect contamination and the quality of the anodized film. However, as yet, this has not been used in practical applications.

Bonding procedure. The main process quality control here is to measure and record the temperature distribution and the temperature pressure cycle. The quality of the final component is assessed from tests on the coupons and other specimens manufactured along with the component.

Final component. In general, nondestructive inspection (NDI) procedures such as ultrasonic (C-scan) and radiography are usually employed. These can detect most significant defects, such as foreign bodies (e.g., separator film), disbonds, delaminations (in cocured joints), and voids. However, at present, there is no NDI procedure capable of detecting weak bonds.

8.3 MECHANICAL JOINTS

General

This section deals primarily with composite-to-metal joints employing mechanical fasteners, particularly bolts. However, it is worth briefly mentioning an important class of joints formed by a wedging action. Joints of this type are employed for various centrifugally loaded applications such as compressor fan blades, as illustrated in Fig. 8.11, where the outward forces develop the flatwise wedging forces. The major advantage of this type of joint is that holes piercing the composite structure (with the associated stress concentrations) are not required. Wedges may be formed by a bonded-on doubler or, more usually (Fig. 8.11) by an insertion of extra material, such as extra graphite/epoxy layers or glass/epoxy layers. Stress concentrations at the grip exit region are minimized by using the softer glass/epoxy material.

A type of joint, intermediate between the wedge and the conventional bolted joint, is the wraparound joint. This involves the use of the tows of fibers wrapped around a loading pin. This configuration has the advantage that the fibers are continuous and, thus, the stress concentrations are relatively small. Bearing is by flatwise compression on the composite and, in this respect, is similar to that encountered in the wedge joint. Joints of this type are employed for helicopter blades, where the fibers forming the cap of the spar form the wrap around root.

General Design Considerations — Fastened Joints

Many of the joint configurations used for bonding (Fig. 8.1) can also be employed for mechanical fastening, since the problems of smoothly transferring load from one adherend to another are quite similar. The main differences are that, in mechanically fastened joints, most of the shear is transmitted through discrete high-modulus regions (the individual fasteners) and the shear loads in the fasteners are reacted out by the compression (bearing) on the faces of holes passing through the joint members. Some of the shear load is also transmitted by friction through the faces of the members. However, although friction is beneficial in improving load transfer in the joint and in reducing bearing loads, friction transfer usually cannot be maintained at a high level during prolonged service due to the loss of clamping pressure resulting from vibration and wear. Further, complications may also arise due to fastener bending, which may occur from the loss of clamping pressure and hole elongation during service. This is a particular problem with composites that are relatively soft and can

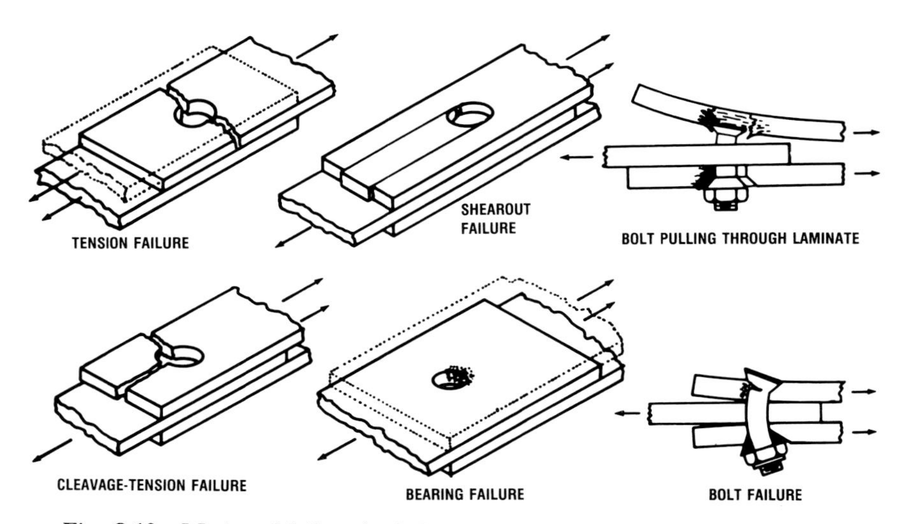

Fig. 8.12 Modes of failure for bolted joints in advanced fiber composites.

result in fatigue failure of the fastener. As in metallic joints, three main modes of failure of the joint members must be considered for design purposes (Fig. 8.12). These are the net section tension or compression, shear, and bearing; however, in practice, mixed modes of failure often occur. The allowable stresses in each of these modes is a function of (1) the geometry of the joint (including the hole size, plate width, and the distance of the hole from the edge of the plate, see Fig. 8.13); (2) the clamping area and pressure—allowing for any countersink; (3) the fiber orientation ply sequence, (4) the moisture content and exposure temperature, and (5) the nature of the stressing (e.g., tension or compression, sustained or cyclic, and any out-of-plane loads causing bending).

Although practical joints are complex, most can be modeled as simple single- or double-lap joints, with single or multiple rows of fasteners. The following discussion on design for graphite/epoxy-to-metal joints considers mainly a double-lap joint (thus avoiding complications caused by bending) with a single-bolt fastener as shown in Fig. 8.13. The ply configuration is taken as the $0°/\pm 45°$ type (with a relatively small amount of 90° plies) commonly employed in aircraft applications. In this section, the various failure modes under externally applied loads will be considered. The effects of thermal expansion mismatch between the composite and metallic component, while possibly a significant factor in the loading of the joint, have apparently not been considered in the literature and are not covered here.

Design Criteria for Failure Under Static Loads

Tensile failure. Estimation of the static tensile strength of a metallic joint is usually based on the net section area. Stress concentration effects are

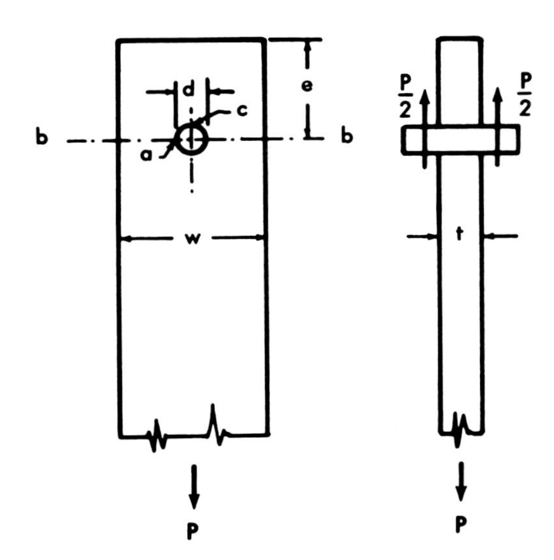

Fig. 8.13 Geometry of a simple double-lap bolted joint.

ignored since it is assumed that they are relieved by local yielding (a dangerous assumption under fatigue loading). Thus, for a joint of the type shown in Fig. 8.13, one can write

$$\sigma_{Tu} = P/(w-d)t \tag{8.8}$$

where σ_{Tu} is the allowable static strength in tension (e.g., the A or B allowable strength for aircraft design) and P the allowable load on the joint.

This situation cannot be assumed for the composite, since general yielding is not possible (although there is some evidence that limited hole softening occurs due to a yielding effect associated with the high interlaminar shear at the edges of the hole). Thus, the simplest relationship that can be assumed is

$$\sigma_{Tu} = K_T P/(w-d)t \tag{8.9}$$

The problem is to determine the stress concentration factor K_T for a loaded hole in a composite plate of a given constitution and configuration.

It is instructive, initially, to consider the K_T for an unloaded hole since this case has been solved analytically. The analysis in Ref. 4 shows that, for an infinite plate, the effective K_T at the edge of the hole at 90° to the applied load is

$$K_T = 1 + \sqrt{\left\{2\left[\sqrt{(E_x/E_y)} - \nu_{xy}\right] + E_x/G_{xy}\right\}} \tag{8.10}$$

where E_x and E_y are the moduli in the direction of the load and normal to the direction of the load, respectively; G_{xy} is the in-plane shear modulus; and ν_{xy} is the major Poisson's ratio.

From this equation, it can be shown that, typically, K_T varies from 7.5 for a 100% 0° laminate to about 1.8 for a 100% $\pm 45°$ laminate and approaches the isotropic value of 3 at about 80% $\pm 45°$. The $\pm 45°$ plies essentially serve to reduce K_T by carrying the load around the hole. It is of interest to note that K_T falls below the isotropic value at 100% $\pm 45°$ plies.

Since the tensile strength of the laminate falls with increasing $\pm 45°$ content, it can be shown[5] that the optimum configuration for the holed laminate is given by the maximum value of σ_{Tu}/K; this occurs at about 50% $\pm 45°$.

These considerations, based on the stress concentration factor at the 90° position are highly simplified because, in general, the peak stress occurs at some other angle; e.g., for a $\pm 45°$ laminate, the peak occurs at 45° to the loading direction where K_T is about 2. Second, the stresses in the individual plies differ according to their orientation, so each ply should be examined separately. Third, biaxial stresses should be considered to obtain the true failure criterion for each layer. Finally, as mentioned earlier, high interlaminar shear stresses at the edges of the hole (dependent on ply constitution and lay-up sequence) result in a local softening that is possibly due to local delaminations, thus reducing the K_T value.

Various procedures have been developed to obtain K_T values for loaded holes, including finite element procedures, semiempirical procedures based on experimentation, and analytical procedures. In general, these are rather complex, especially for the purpose of the present discussion. However, from experimental studies,[5] it was shown that Eq. (8.11) gives a reasonable estimate of the effective K_T for a loaded hole, at least for one representative situation.

Compression failure. It will be shown later that fatigue must be considered for the design of joints subjected to compression loading, unlike tension loading where static strength is the design criterion. Although, for an unfilled hole, stress concentration factors similar to those discussed above must be considered for compression, the presence of a fastener stabilizes the surrounding material against compression failure.

The mode of failure under compression loading is dominated by the shear properties of the matrix, because the effective shear modulus of the matrix is a major factor determining the stability of the fibers against buckling. The effective shear modulus of the matrix decreases with increasing temperature (with the maximum service temperature usually about 120°C) and also with increasing absorbed moisture. Moisture dissolved in the resin matrix (up to a maximum of about 1.5% of the laminate weight) acts as a plasticizer. The compression strength for the hot/wet condition can be as low as 50% of the room temperature/dry condition.

Shear failure. Estimation of shear or tear-out failure load for a metallic joint is usually based on the following equation:

$$\tau_u = P/2et \tag{8.11}$$

where τ_u is the allowable shear stress and the rest of the notation is as indicated in Fig. 8.13. There is experimental evidence[5] that this simple relationship also holds for graphite/epoxy laminates containing a significant fraction of $\pm 45°$ plies, which implies that there is no stress concentration for this type of loading.

In contrast, significant stress concentrations appear to arise for an all 0° laminate (not a practical configuration for a bolted joint). With an all 0° laminate, splitting is the preferred mode of failure. Fairly elaborate precautions are required to inhibit this mode in order to allow for measurement of the shear strength. When this is done, it is found that the value obtained for τ_u falls well below the apparent in-plane shear strength and the interlaminar shear strength. Splitting is suppressed for laminates containing plies at significant angles to the 0° direction (e.g., $\pm 45°$) since these provide transverse strengthening. Experiments[5] show that an optimum value of strength occurs for the 0, $\pm 45°$ family of laminates, again at a combination of 50% 0° plies and 50% $\pm 45°$ plies.

Bearing failure. Bearing loads for a metallic joint are usually estimated from the simple relationship

$$\sigma_{Bu} = P/dt \tag{8.12}$$

where σ_{Bu} is the allowable bearing stress and the other notation is as indicated in Fig. 8.13. Failure in bearing in a metallic joint occurs by compression yielding of the material at the compression end of the hole. In general, the bearing load is the highest load that can be carried by the joint if e and w are large enough to avoid tension or shear-out failures. In a composite, bearing failures occur by local buckling and kinking of the fibers and subsequent crushing of the matrix. However, according to Ref. 5, Eq. (8.12) is applicable.

As a result of this failure mode, σ_{Bu} is quite strongly dependent on the lateral constraint (clamping pressure) provided by the fastener. Experimental studies[5] have shown that an improvement in bearing load of 60–170% can be obtained over pin loading, with a fastener clamping pressure of about 22 MPa; above this pressure no further improvement is obtained. Since clamping pressure plays an important role in the bearing strength of composites, it will be important (particularly if the design is based on constrained bearing) to ensure that this pressure is maintained during the service life.

Bearing stresses initially increase with increasing proportions of 0° plies, since these are most efficient in carrying bearing loads. However, with laminates having above about 60% 0° plies, failure occurs by splitting, since the 0° plies have a low transverse strength. The optimum once again appears to occur at about 50% 0° and 50% $\pm 45°$. It is found that the bearing strength is further improved as the ply sequence is made more homogeneous (dispersion of 0° and $\pm 45°$ plies). Theoretically, it is well known that interply shear stresses are reduced as the laminate becomes more homogeneous.

The bearing strength may also be expected to be dependent on the moisture content of the resin matrix and the exposure temperature. While, in general, bearing strength probably does decrease with the matrix moisture content and test temperature, these adverse effects are considerably reduced if a high level of local lateral constraint is present. Further, in practice, some hole softening leads to better contact of the fastener with the hole and so reduces high local loads. Thus, in general, environmental effects do not appear to affect bearing strength significantly.

Influence of Fatigue Loading

Open holes. In order to discuss the effect of fatigue loading on a graphite/epoxy laminate with an open hole, it is necessary to consider the mechanism of failure in a little detail; Ref. 6 includes several papers of relevance here. Under tension or compression-dominated fatigue loading, the sequence of failure begins in regions of high stress concentration, with matrix cracking and local debonding of fibers from the matrix. The localized damage accumulates until it results in more extensive intralaminar cracking. This reduces the in-plane stress concentration, but increases interlaminar shear stresses, eventually producing delaminations. The rate and extent of degradation depend on the interply shear stresses, which are strongly dependent on the ply configuration. More homogeneous configurations generally produce lower interply stresses.

At this stage, tension and compression behavior differs greatly. The effect of the degradation process under fatigue is to reduce the tensile stress concentration; therefore, at least in the early stages, the residual strength may show a significant increase. In any event, within the allowable static tensile strength range, the effect of fatigue on residual strength is usually not detrimental, unless the delamination damage is very severe. Thus, for design purposes, it is necessary to consider only the static strength for laminates subjected to tensile loading.

Under compression loading, although the compression stress concentrations around the hole are similarly reduced, the loss in section stiffness due to delamination can lead to compression or buckling failure of the remaining sound material.[7] Thus, residual compression strength falls as a function of fatigue loading; consequently, fatigue must be considered in the design of laminates subjected to compression loading. The extent to which large-scale delaminations can grow under fatigue loading and the critical size of delaminations for failure under limit loads are subjects of current research.

The compression strength of the composite is, as mentioned previously, dominated by the mechanical properties of the matrix. As a consequence, fatigue properties under compression-dominated loading are sensitive to the moisture content of the matrix and the test temperature.

Filled holes/compression loading. The above considerations may also be relevant to laminates with a filled (or loaded) hole, even if the fastener is initially a close fit in the hole. This is because partial bearing failure may occur under fatigue loading, allowing fastener movement and thus spoiling the close fit. In practice, hole enlargement and loss in residual strength during fatigue loading do not appear to be problems (residual strength may even increase), if load reversal does not occur, e.g., for $R = -\infty$ (zero compression). However, loss of residual strength is marked at $R = -1$ (tension compression). The major problem at $R = -1$ is that of fastener failure due to the fastener bending consequent on the hole enlargement. Thus, the actual situation under fatigue loading is quite complex. A possible sequence under $R = -1$ is (1) loss of clamping pressure under fatigue loading, (2) partial bearing failure and hole enlargement resulting from loss of clamping pressure, (3) fastener failure, and (4) final failure due to the overloading of adjacent fasteners and holes.

To allow for the presence of delaminations, which may arise due to fatigue damage around stress concentrations or from low-energy impact damage, the design allowable for compression-dominated cyclic loading is usually given as an allowable strain in the material away from the region of the stress concentrator. The advantage of this approach is generally that no further specification of laminate stiffness is required. Most aircraft manufacturers take the allowable value to be between -3000 and -4000 microstrain.

Materials and Processing Considerations

Hole strengthening procedures. Several procedures may be employed to improve the strength of the composite under loaded hole conditions. Most of these procedures are based on the incorporation of extra layers into the laminate in the holed region. However, although the composites lend themselves well to modifications of this nature, manufacturing costs are usually greatly increased. Consequently, such procedures are generally confined to critical applications such as highly loaded lugs.

The stress concentration at the edges of a loaded hole in graphite/epoxy can be reduced significantly, either by local reinforcement with a stiffer fiber such as boron or by local softening created by local reinforcement with a low-modulus fiber such as glass or aramid; these procedures are also employed to improve the damage tolerance of the composites. An effective alternative means of softening the holed area is by incorporation of extra $\pm 45°$ graphite/epoxy plies; an alternative means of reinforcement is by incorporation of layers of titanium alloy sheet. The last two methods are also very effective in improving the bearing and shear strength, both by providing additional reinforcement (the titanium alloy is particularly effective in improving bearing strength) and by increasing the area.

A simpler and less costly approach is to reinforce the hole with an externally bonded doubler, made either of a composite (glass, aramid, or graphite) or titanium alloy. The doubler must be appropriately scarfed to minimize shear and peel stresses in the adhesive. In an experimental study in which the weight and thickness of some of these concepts were compared for a given load-carrying capacity, it was found that use of extra $\pm 45°$ graphite/epoxy plies was the lightest solution and use of titanium alloy interleaves the thinnest. However, the use of titanium created considerable manufacturing difficulty because of the bonding pretreatment required and the subsequent difficulty in forming the holes (titanium tends to cause delamination of the material around the hole).

Hole formation practice. Hole drilling in graphite/epoxy, using tungsten carbide tipped drills, causes no particular problem provided simple precautions are taken to avoid delamination. In particular, care must be taken to support the laminate during drilling by sandwiching it either between scrap material or in a drilling guide jig. Delamination on the exit side of the drill can also be prevented by coating the composite on this side with a layer of film adhesive. It is also advantageous to employ a pressure-controlled drill and a fairly slow feed rate. Under mass production conditions, some minor delamination damage is inevitable. Provided the delamination is not too extensive, the damage can be repaired by the local injection of an epoxy resin.

Although very good tolerances can be maintained on holes in graphite/epoxy, interference fit fastening is generally (although not universally) avoided, since excessive interference can result in delamination damage; significant stressing of the hole may also arise subsequently at elevated temperature resulting from differential expansion of the fastener and laminate. Generally, clearance holes with a tolerance of 0.05–0.1 mm are used in most aircraft applications.

For applications where the fastener heads are required to be flush with the surface of the components, countersink depths are limited to avoid the formation of knife edge bearing surfaces, which are very fragile in composite materials.

Fastening procedures. The fastener spacing and edge distance employed depend on factors such as the type of fastener used, the load transfer requirements, and the ply configuration. Generally, the minimum edge distance ranges from two to three times the fastener diameter and the minimum allowable fastener spacing ranges from three to four fastener diameters.

For most applications, the recommended[5] clamping pressure σ_z is about 20 MPa, the required torque T may be calculated from $\sigma_z = 1.658/d^3$ where a washer of diameter $2.2d$ is assumed.

Corrosion prevention. Graphite/epoxy materials are electrically conducting and cathodic with respect to most structural metals. Thus, to avoid the danger of galvanic corrosion of the metal side of the joint, special precautions are required. Glass, aramid, and boron fiber reinforced plastics are nonconducting and thus pose no corrosion problems.

In general, fasteners made from aluminum alloys or steel are avoided unless they can be insulated from the graphite/epoxy composite. The preferred fastener material, particularly for bolts and lock pins, is titanium alloy, although stainless steel is also considered to be suitable. Where the titanium alloy fastener comes into contact with the aluminum alloy side of the joint, an aluminum pigmented coating may be used for corrosion compatibility. Corrosion resistant steel nuts and washers are used, which may be cadmium plated where they come into contact with aluminum structure.

In areas where graphite/epoxy and aluminum alloy may come into contact, an insulating layer of glass/epoxy or aramid/epoxy is employed. This is usually cocured onto the surface of the graphite/epoxy laminate during manufacture. In some cases, the insulating layer may also be employed on the outside of the component to allow use of aluminum fasteners.

Fasteners are usually wet set in a corrosion-inhibiting sealant, such as a chromate-pigmented polysulphide rubber. However, this precaution is not necessary for titanium, unless it is required for sealing areas containing fuel.

References

[1]Kinloch, A. J., *Developments in Adhesive-2*, Applied Science Publishers, London, 1981.

[2]Hart-Smith, L. J., "Adhesive-Bonded Double-Lap Joints," NASA CR-112235, 1973.

[3]Hart-Smith, L. J., "Analysis and Design of Advanced Composite Bonded Joints," NASA CR-2218, 1974.

[4]Lekhnitski, S. G., *Anisotropic Plates*, 2nd ed., Gordon and Breach, New York, 1968, p. 171.

[5]Collings, T. A., "The Strength of Bolted Joints in Multi-Directional CFRP Laminates," *Composites*, Vol. 8, No. 1, 1977, pp. 43-55.

[6]*Fatigue of Filamentary Composites*, ASTM STP 636, 1977.

[7]Saff, C. R., "Compression Fatigue Life Prediction Methodology for Composite Structures," NADC-78203-60, 1980.

9. ENVIRONMENTAL EFFECTS AND DURABILITY

9.1 INTRODUCTION

In this chapter, key factors influencing the performance of composite aircraft structures in service will be discussed. These include environmental effects, especially the effect of moisture absorption from the atmosphere taken in conjunction with elevated temperatures such as are encountered by an aircraft structure in high-speed flight, and the effect of cyclic (fatigue) loads.

Bearing in mind that metal aircraft structures have been in widespread use for almost half a century and that significant uncertainties about metals still exist in the general areas of environmental effects and fatigue performance, it can hardly be considered surprising that significant uncertainties exist in these same areas for composite aircraft structures where there is so much less experience. Another aspect that compounds the uncertainties for composites is the presence of additional parameters, which have no analog for metals and which can affect performance, e.g., the ply orientations and stacking sequence for a laminate. It is usual practice, when assessing environmental effects and fatigue performance for design purposes, to establish a relatively large data base for the particular laminate patterns it is intended to employ.

For further discussions of environmental effects and fatigue, see, for example, Refs. 1–4.

9.2 MOISTURE ABSORPTION

The matrix materials, such as epoxy resins, that are commonly used for present-day composites absorb moisture from the atmosphere by what is essentially a diffusion process. Under ambient temperature conditions, the rate of diffusion is quite slow: it takes times of the order of months for a laminate kept in a humid atmosphere to achieve an equilibrium moisture distribution. The reverse process, called "desorption," also occurs when a laminate containing moisture undergoes long-term exposure in a dry atmosphere it will give up its moisture also by a diffusion process. The amount of moisture in a laminate is generally expressed in terms of the percentage increase in the laminate weight; for graphite/epoxy laminates exposed for long times in humid atmospheres, weight increases of the order of 1% are encountered.

Since, as will be seen later, the presence of moisture in a laminate can significantly affect certain of its structural properties, considerable attention has been devoted to establishing theoretical procedures for predicting the moisture content. These can be illustrated by reference to the following example, which, although a substantial simplification of the situation for an aircraft in service where the environmental conditions vary, is nevertheless representative of the situation encountered in the laboratory in connection with the moisture conditioning of test specimens. Consider, therefore, an initially dry laminate immersed in a constant humid atmosphere (Fig. 9.1). It is assumed that moisture enters the laminate only through its flat faces. (When conditioning test specimens, this is often ensured by protecting the side edges of the laminate with some material that is impervious to moisture.) If z denotes the coordinate in the thickness direction, here measured from one face of the laminate, and t the time, then it has been shown (see, for example, Refs. 5–8) that, at least to a good approximation, the ingress of moisture proceeds in accordance with Fick's law, namely,

$$\frac{\partial c}{\partial t} = D\frac{\partial^2 c}{\partial z^2} \tag{9.1}$$

where $c(z, t)$ denotes the moisture concentration in the laminate at depth z at time t and D is the diffusion constant. Also, there are initial conditions that, for the present case of an initially dry laminate, become simply

$$c(z,0) = 0, \qquad 0 < z < h \tag{9.2}$$

where h is the thickness of the laminate. Finally, there are the boundary conditions, which here are

$$c(0, t) = c(h, t) = c_0 \tag{9.3}$$

where c_0 is a constant that can be related to the humidity of the atmosphere in which the laminate is immersed.

Equation (9.1) is a very well-known one and a variety of analytic and numerical solution methods are available.[9] For the simple initial and boundary conditions (9.2) and (9.3), the method of "separation of variables" can be used to obtain the following solution:

$$\frac{c(x, t)}{c_0} = 1 - \frac{4}{\pi}\sum_{n=0,1,\ldots}\frac{\sin[(2n+1)\pi z/h]}{(2n+1)}\exp\left[-D(2n+1)^2\frac{\pi^2 t}{h^2}\right] \tag{9.4}$$

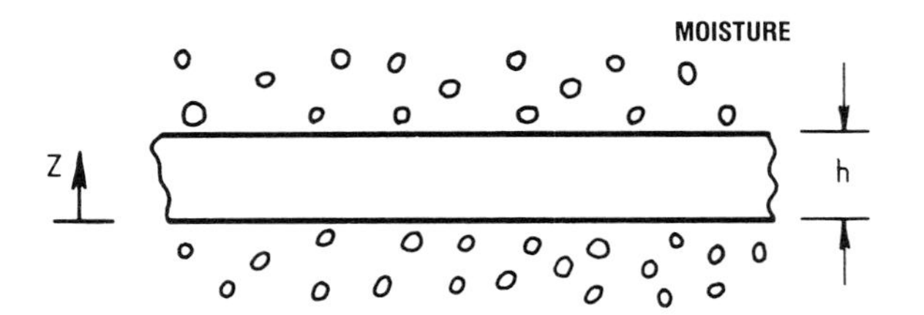

Fig. 9.1 Moisture absorption by a laminate.

Equation (9.4) gives the concentration of moisture at various depths in a laminate at various times; the general nature of the results is as shown in Fig. 9.2, where it can be seen that, for sufficiently long times, the moisture concentration approaches a uniform value across the depth of the laminate ("saturation"). The total mass of moisture m in a laminate can be obtained by integrating Eq. (9.4) across the depth of the laminate, i.e.,

$$m = \int_0^h c(z, t)\,\mathrm{d}z \tag{9.5}$$

Performing the integration gives

$$\frac{m}{m_0} = 1 - \left(\frac{8}{\pi^2}\right) \sum_{n=0,1,\ldots} \frac{\exp\left[-D(2n+1)^2\pi^2 t/h^2\right]}{(2n+1)^2} \tag{9.6}$$

where $m_0 = c_0 h$.

For sufficiently small t, it can be shown that Eq. (9.6) is closely approximated by the following simple formula:

$$m/m_0 = (4/h)(Dt/\pi)^{\frac{1}{2}} \tag{9.7}$$

Thus, in its initial stages, the moisture absorption proceeds at a rate proportional to the square root of the time. For this reason, in graphical presentations, it is usual to plot the moisture absorption against the square

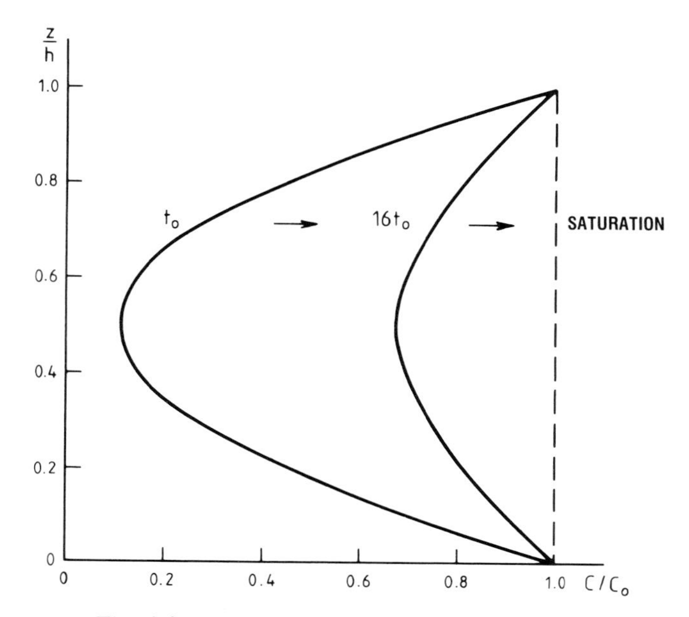

Fig. 9.2 Moisture/time profiles in a laminate.

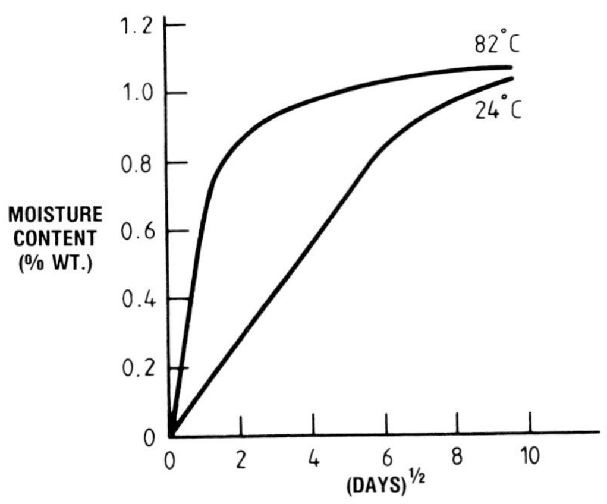

Fig. 9.3 Effect of temperature on the rate of moisture absorption (eight-ply graphite/epoxy laminate, 75% relative humidity).

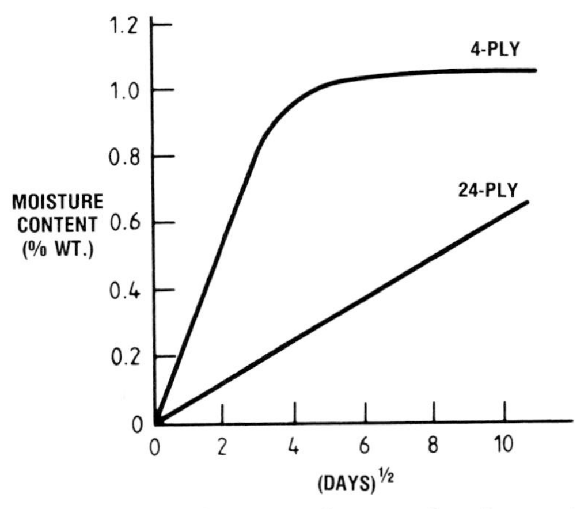

Fig. 9.4 Effect of laminate thickness on the rate of moisture absorption (graphite/epoxy laminate, 24°C, 75% relative humidity).

root of the time; when this is done, the initial part of the curve is a straight line and the diffusion constant D can be determined experimentally by measurements of the slope of this line.

Two general features of moisture absorption in laminates are:

(1) The higher the ambient temperature, the more rapidly the moisture is absorbed; the diffusion constant is, in fact, quite strongly dependent on the temperature.

(2) The thicker the laminate, the longer it takes for the saturation condition to be achieved.

These two features are illustrated in Figs. 9.3 and 9.4 where the data have been taken from Ref. 6.

For a laminate forming part of an aircraft in service it is necessary to replace the simple boundary conditions [Eq. (9.3)] by ones that represent the climatic variations in humidity; it is also necessary to allow for the effects of temperature variations on the diffusion constant. A program studying moisture absorption in the boron/epoxy skins of the F-15 horizontal tail is described in Ref. 10. There, using the general methodology outlined above, predictions have been made of the moisture absorption over 25 years of service life of the F-15. Certain horizontal tails will be periodically removed from service, their moisture content measured, and the measured values compared with the theoretical predictions. Values measured after 5 years of service compared well with the theoretical predictions; at the time of measurement, a typical value of the moisture absorption was 0.6%, although higher values (around 0.8%) were predicted at various intermediate times. Another extensive program on the moisture absorption occurring in service involves graphite/epoxy spoilers on Boeing 737 aircraft operating in various parts of the world.[11] A large number of spoilers are again being regularly removed from service and the change in weight of the laminate specimens is measured as the specimens are dried out; it appears that weight changes of the order of 0.8% are being measured.

One effect of moisture absorption is that stresses can be induced due to swelling,[8] especially when there is a moisture gradient across the thickness of a laminate (as in Fig. 9.2), so that there is unequal swelling of different plies. However, a more important effect of moisture absorption is described in the next section.

9.3 EFFECT OF MOISTURE AND TEMPERATURE ON STRUCTURAL PERFORMANCE

It transpires that the main effect of moisture on structural performance comes not from the moisture alone, but rather when it is taken in combination with elevated temperatures such as occur in high-speed flight. Composite components on a supersonic fighter aircraft might be expected to experience temperatures in the approximate range -50 to 125°C depending on speed and altitude; it is the high end of this range that is seen as being of most importance.

As has already been mentioned in Chap. 4, the effect of absorbed moisture on a resin is to reduce its glass transition temperature. (Broadly, the glass transition temperature of a polymer can be defined as that temperature above which the polymer is soft and below which it is hard.) When a fiber/epoxy composite is in a "wet" condition (having a moisture content around, say, 1.2%, then the glass transition temperature of the epoxy matrix may be reduced by around 40–50°C; see Ref. 5. Under these circumstances, the maximum temperature seen in service may be approaching the glass transition temperature of the composite matrix. This is where the main potential for reductions in structural performance arises. However, by no means are all of the structural properties of a composite reduced significantly from this cause. As would be expected from the above argument, it is only those properties strongly dependent on the matrix performance that are affected.

According to Ref. 7, the tensile strength and extensional stiffness of fiber-dominated laminates (i.e., laminates with a significant proportion of fibers in the principal stress directions) are little affected by temperature over the range -50 to 125°C regardless of the moisture content, except possibly for a slight decrease in tensile strength when the moisture content exceeds 1% and the temperature exceeds 110°C. (Actually, according to Ref. 12, it is the dry, cold (-50°C) condition that is often critical for the tension strength of composite structures.) The situation is different for matrix-dominated laminates, where substantial reductions in tensile strength and stiffness can occur. (However, such laminates are not generally used for structural applications, although an ever-present problem in composite design is the possibility of hidden load paths, i.e., the existence of loads acting in directions that were not anticipated during the design and where, in consequence, no fibers may have been appropriately oriented.)

It is the compression properties of composite laminates that are most susceptible to moisture/temperature effects. In a general sense, this might be expected because of the nature of fiber composites. While individual fibers can carry tensile loads on their own, it is necessary that they be bonded into a structural entity with some form of support against buckling if they are to carry compression loads; it is the matrix material that provides this bonding and support and, if the general solidity of the matrix is impaired, some degradation of a composite's compression properties becomes likely.

A summary of compression strength tests on 48-ply laminates of graphite/epoxy AS/3501-6 with a lay-up comprising 42% 0s, 50% $\pm$45s, 8% 90s is given in Ref. 12; there, the compressive strength at 120°C with 0.6% moisture absorption appears to be approximately 33% less than the strength in the room temperature dry condition. Broadly similar "representative data" are cited in Ref. 13, although there, while the laminates, temperature, and reduction in compressive strength are much the same as in Ref. 12, a rather larger moisture content (1%) is quoted. Considerable effort is now being devoted to the development of resin systems that are less prone to moisture absorption than the epoxies; the most promising of these to date seems to be the PEEK (polyether-ether-ketone) system.

9.4 OTHER ENVIRONMENTAL EFFECTS

The external surfaces of graphite/epoxy aircraft structures are generally painted to prevent any long-term degradation due to ultraviolet radiation; the paint also serves to prevent the otherwise high absorption of thermal (e.g., solar) radiation by a "black body." (However, the paint has no significant effect in reducing moisture absorption, although efforts are being made to develop moisture-impervious coatings.)

With regard to the susceptibility of graphite/epoxy to attack by fluids such as fuel, hydraulic fluid, and paint strippers, which are commonly encountered in aircraft usage, it appears that only paint strippers have a significantly deleterious effect. (However, it might be noted that some other resin systems, e.g., polysulfones, have shown a susceptibility to attack by hydraulic fluids.)

9.5 FATIGUE OF COMPOSITES IN NORMAL ENVIRONMENT

Turning now to the fatigue of composites, this will be discussed first for a normal environment (i.e., room temperature and moderate humidity). It needs to be said at the outset that the fatigue of composites, especially advanced composites, is different in many respects from the fatigue of metals. This remark applies not only to the factors influencing fatigue performance (such as type of loading and notch sensitivity) but also to the way in which fatigue damage manifests itself. On this latter point, while in a metal, fatigue damage eventually appears as a clearly defined crack, in a composite, the situation is much more complex. Fatigue damage can take the form of some or all of the following: matrix cracks, fiber/matrix debonds, fiber fractures, and delamination growth (i.e., interply debonds).

For relatively recent reviews of composite fatigue, see, for example, Refs. 14 and 15.

Tension-Dominated Fatigue

Plain laminates. Here attention is confined to plain laminates (i.e., without stress concentrators) that are subject to tension-tension cycling, so that the fatigue stress ratio R (defined as the ratio of the minimum to the maximum cyclic stress) is non-negative. Under these circumstances, it has been established that fiber-dominated laminates made of advanced composite materials have a much better fatigue performance than structural metals. For example, according to Ref. 16, unidirectional graphite/epoxy laminates exhibit almost no sign of a conventional S-N curve. Instead, a wide scatter of lifetimes ranging out to 10^7 cycles is found, even when the maximum cyclic stress is around 80% of the ultimate tensile stress (UTS); if the maximum cyclic stress is less than 70% of the UTS, then almost no failures are observed in cycling with $R = 0$ out to 10^7 cycles. This feature is illustrated in Fig. 9.5, where the data have been taken from Ref. 14. Multidirectional laminates do exhibit some sensitivity to fatigue loadings and an S-N curve can be obtained. However, as long as the laminate is fiber dominated, the fatigue performance remains very good. For example, Ref. 17 gives results for $[(\pm 45)_2/0_2]_s$ graphite/epoxy laminates that sustained 10^6 cycles with the maximum cyclic stress 80% of the UTS.

The situation is much the same for boron composites (see p. 404 of Ref. 14) and, apparently, for aramid composites.[4] The good tension fatigue performance of advanced composites is generally attributed to their high Young's modulus; the fatigue performance of the lower-modulus glass composites is distinctly poorer,[4] although still comparable with that of aluminum alloys.

Laminates with stress concentrators. In structural metals, small holes or other forms of stress concentrators generally have little effect on the static strength, but have a marked effect in reducing fatigue life. The situation for composites seems to be the complete reverse. Because fiber composites are fairly brittle materials, the presence of stress concentrators in them can lead

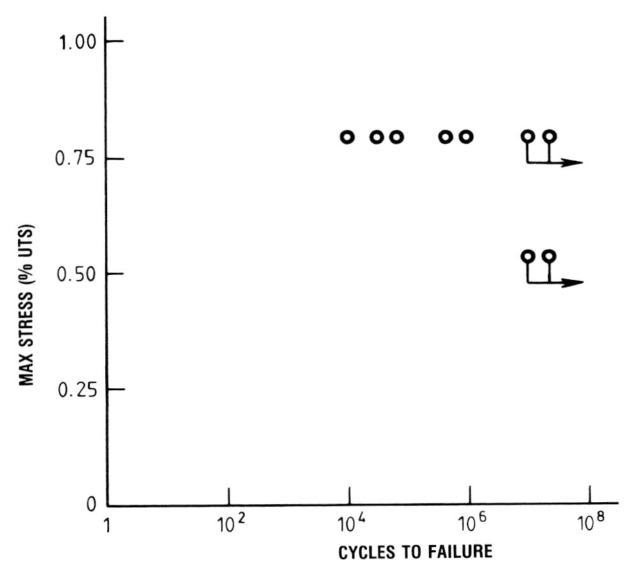

Fig. 9.5 S-N data for unidirectional graphite/epoxy laminate (UTS 1345 MPa, $R = 0$).

to a significant reduction in their static strength (although the reduction is generally not as severe as would be predicted by the theoretical stress concentration factors discussed in Chap. 7). However, subsequent cycling often leads to no further degradation in strength; indeed, the residual strength after cycling can exceed the original strength (see p. 408 of Ref. 14). This rather paradoxical situation apparently comes about because of the following effect. Although the fatigue cycling can initially cause a limited amount of damage around the stress concentrator, the result of this damage can be to redistribute the stresses (especially in laminates with various ply orientations) so that the effect of the stress concentrator is reduced. Using laminates of the type described earlier, Bevin[17] found that the presence of a 2 mm drilled hole in a 25 mm wide specimen could reduce the static strength of the specimen by up to 50%, but that then lives of 10^6 cycles, with the maximum stress in the cycle virtually equal to the (reduced) static strength, could be obtained; the residual tensile strength was also very significantly increased. Sturgeon[16] has given generally similar results, except that lesser static strength reductions (around 24%) occurred.

Compression-Dominated Fatigue

For metals, fatigue is governed largely by the tensile stresses; once again, the situation seems to be quite the reverse for composites where, although there is some contradictory evidence, it appears that compressive stresses play the key role. For example, Rosenfeld and Huang[18] carried out tension-compression ($R = -1$) and zero-compression ($R = -\infty$) fatigue tests on graphite/epoxy laminates ($0/\pm 45$) with holes and concluded that compres-

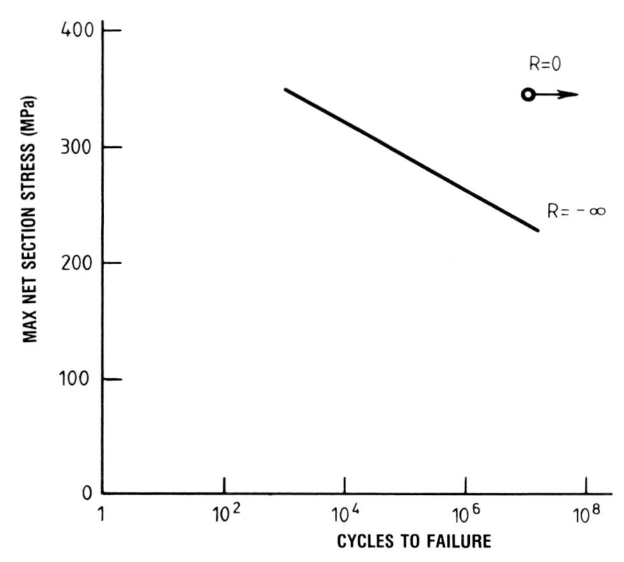

Fig. 9.6 S-N data for graphite/epoxy (0/ ± 45) laminate under tension ($R = 0$) and compression ($R = -\infty$) fatigue. (Taken from Ref. 18.)

sive loads in fatigue produce a significant reduction in fatigue life when compared with results for tension-tension loading; see Fig. 9.6. They describe the mechanism of failure for compression fatigue as a progressive local fatigue failure of the matrix near a stress riser, thus causing fiber split, progressive delamination, and fiber buckling, which then precipitates laminate failure. Similarly, Ryder and Walker[19] concluded that, for graphite/epoxy, compression markedly reduces fatigue life, with a greater reduction corresponding to a larger compression load excursion.

9.6 FURTHER COMMENTS ON COMPOSITE FATIGUE

Environmental Effects

Studies of the influence of moisture and temperature on the fatigue performance of composites are still in their relative infancy, and many uncertainties exist. It seems to be implied by data summarized in Ref. 12, relating to compression-dominated fatigue of graphite/epoxy laminates with moisture levels of 0.6–1.5%, that, as long as environmental effects are allowed for in the static design, a satisfactory fatigue performance can be expected. However, one area where substantial concern has been expressed is the effect of "thermal spikes"; these are relatively large increases in temperature occurring over short time periods (at rates approaching, say, 1°C/s) corresponding to an aircraft accelerating rapidly to supersonic flight. Especially for "wet" laminates, there is the possibility of damage to the matrix occurring, leading to a degradation in fatigue life and/or residual strength under compression loadings. This type of effect has been observed in a test program on graphite/epoxy structures.[20]

Cumulative Damage

Although its shortcomings are realized, Miner's rule for cumulative damage in metal fatigue generally provides results that are, at least, of the right order of magnitude and have no consistent bias toward either conservatism or nonconservatism.[15] There are some very preliminary indications[15,18] that, for composites, the rule may be markedly nonconservative.

Statistical Aspects

The indications from specimen tests are that the scatter in the fatigue life of composites is significantly greater than that of metals. (At this stage, of course, there are little statistical data on the fatigue of composite structures, but in reading the following it should be borne in mind that as a general rule the scatter in life of built-up structures is usually significantly less than that for simple specimens.) The Weibull probability distribution is the one that seems to be most favored for representing the variation in composite fatigue life.[19,21] According to Ref. 19, whereas the number of specimens used for defining the scatter in composite fatigue life is often taken to be around 20, the extent of the scatter is such that this cannot be used to justify a survival rate greater than 95%; it is stated that, to justify a survival rate greater than 99%, it is necessary to test at least 100 specimens.

9.7 FINAL REMARKS

As was stated at the beginning of this chapter, there are currently many uncertainties in all of the topics that have been discussed. However, while various laboratory programs have identified several areas of potential concern, as far as is known no major problems have so far arisen due either to environmental effects or fatigue in the composite components of aircraft in service. In the case of boron/epoxy, some such components have now been in service for almost 15 years; graphite/epoxy components have been in service for a lesser time. Generally, what are considered to be conservative design procedures have been used for current composite aircraft structures.

References

[1] *Environmental Effects on Advanced Composite Materials*, ASTM STP 602, 1976.

[2] *Advanced Composite Materials—Environmental Effects*, ASTM STP 658, 1978.

[3] Reifsnider, K. L. and Lauraitis, K. N. (eds.), *Fatigue of Filamentary Composite Materials*, ASTM STP 636, 1977.

[4] Agarwal, B. D. and Broutman, L. J., *Analysis and Performance of Fiber Composites*, John Wiley & Sons, New York, 1980, Chap. 7.

[5]Browning, C. E., Husman, G. E., and Whitney, J. M., "Moisture Effects in Epoxy Matrix Composites," *Composite Materials: Testing and Design (Fourth Conference)*, ASTM STP 617, 1977, pp. 481–496.

[6]Shirrell, C. D. and Halpin, J., "Moisture Absorption and Desorption in Epoxy Composite Laminates," *Composite Materials: Testing and Design (Fourth Conference)*, ASTM STP 617, 1977, pp. 514–528.

[7]Springer, G. S., "Environmental Effects on Epoxy Matrix Composites," *Composite Materials: Testing and Design (Fifth Conference)*, ASTM STP 674, 1979, pp. 291–312.

[8]Tsai, S. W. and Hahn, H. T., *Introduction to Composite Materials*, Technomic Publishing Co., Westport, CT, 1980, Chap. 8.

[9]Crank, J., *The Mathematics of Diffusion*, Oxford University Press, London, 1956.

[10]Hinkle, T. V., "Effect of Service Environment on F-15 Boron/Epoxy Stabilator," AFFDL-TR-79-3072, 1979.

[11]Stoecklin, R. L., "Commercial Service Experience of a Graphite-Epoxy Flight Control Component," *Proceedings of Ninth National SAMPE Technical Conference*, 1977, pp. 289–296.

[12]Weinberger, R. A., Somoroff, A. R., and Riley, B. L., "U.S. Navy Certification of Composite Wings for the F-18 and Advanced Harrier Aircraft," AGARD-R-660, 1977, pp. 1–12.

[13]Guyett, P. R. and Cardrick, A. W., "The Certification of Composite Airframe Structures," *Aeronautical Journal*, Vol. 84, 1980, pp. 188–203.

[14]Hahn, H. T., "Fatigue Behavior and Life Prediction of Composite Laminates," *Composite Materials: Testing and Design (Fifth Conference)*, ASTM STP 674, 1979, pp. 383–417.

[15]Heath-Smith, J. R., "Fatigue of Structural Elements in Carbon Fibre Composite—Present Indications and Future Research," Royal Aircraft Establishment, TR 79085, 1979.

[16]Sturgeon, J. B., "Fatigue of Multi-Directional Carbon Fibre Reinforced Plastics," *Composites*, Vol. 8, 1977, pp. 221–226.

[17]Bevan, L. G., "Axial and Short Beam Shear Fatigue Properties of CFRP Laminates," *Composites*, Vol. 8, 1977, pp. 227–232.

[18]Rosenfeld, M. S. and Huang, S. L., "Fatigue Characteristics of Graphite/Epoxy Laminates under Compression Loading," *Journal of Aircraft*, Vol. 15, 1978, pp. 264–268.

[19]Ryder, J. T. and Walker, E. K., "Effect on Fatigue Properties of a Quasi-Isotropic Graphite/Epoxy Composite," *Fatigue of Filamentary Composite Materials*, ASTM STP 636, 1977, pp. 3–26.

[20]Wolff, R. V. and Wilkins, D. J., "Durability Evaluation of Highly Stressed Wing Box Structures," *Fibrous Composites in Structural Design*, edited by E. M. Lenoe et al., Plenum Press, New York, 1980, pp. 761–769.

[21]Phoenix, S. L., "Statistical Aspects of Failure of Fibrous Materials," *Composite Materials: Testing and Design (Fifth Conference)*, ASTM STP 674, 1979, pp. 455–483.

10. DAMAGE TOLERANCE OF FIBER COMPOSITE LAMINATES

10.1 INTRODUCTION

The term "damage tolerance" is used to describe a design philosophy for military aircraft whereby a component is designed such that structural integrity is maintained while a defect of a given size is present in the structure. Modern military aircraft made of metallic materials are designed on this basis, using fracture mechanics to predict the size of a tolerable flaw under the applied loads.

With high-performance composites, the field of damage-tolerant design is complex, due to the inhomogeneous nature of the material and the failure modes, which differ significantly from those in metals. Composites exhibit near-linear stress-strain characteristics up to failure, while most metals display some ductile deformation. Thus, composites are less tolerant of overload. In fatigue, composites again differ from metals in that metals are sensitive to tension-dominated fatigue loading, whereas composites generally exhibit good resistance to tension fatigue. Composites are, however, susceptible to local delaminations that may grow under compression fatigue.

A further complication arises in the manufacturing of composites. Because of the multiphase nature of the material and the processes used in its manufacture, a substantially higher number of defects may exist in a composite component than would occur in a metallic component. Such defects also need to be considered in the damage tolerance assessment of a component.

10.2 NATURE OF DEFECTS

Manufacturing Defects

Defects in composites due to manufacture are usually of two forms: those produced during the preparation and production of the composite and those produced during machining, processing, and assembly of the final component. Typical production defects were shown in Table 6.2. Assembly defects usually arise as a result of damage to the final product by such occurrences as scratches, gouges, impact delamination, fiber breakage, incorrect drilling of holes, overtightened fasteners, etc. In any assessment of the need to repair or reject components, the location and size of the defect, the type of defect, the load spectrum anticipated for the component, and the criticality of the component need to be considered.

Impact Damage

The type of damage resulting from impact on composites depends on the energy level involved in the impact (see Fig. 10.1). High-energy impact, such as ballistic damage, results in through-penetration with perhaps some minor local delaminations. Lower energy level impact, which does not produce penetration, may result in some local damage on the impact zone together with delaminations within the structure and fiber fracture on the back face. Internal delaminations with little, if any, visible surface damage may result from low-energy impact.

Studies[1] on the high-energy ballistic damage problem have shown that, for perpendicular impact on boron/epoxy panels, the damage zone size on the entry face is independent of laminate thickness, while the exit face damage zone has been shown to increase linearly with panel thickness. Other work[2] has shown that the panel thickness has little effect on residual strength. Also, specimens subjected to tension-tension fatigue after ballistic damage showed little variation in residual strength. It was also shown that projectile velocity had no significant effect on residual strength. It was concluded that,

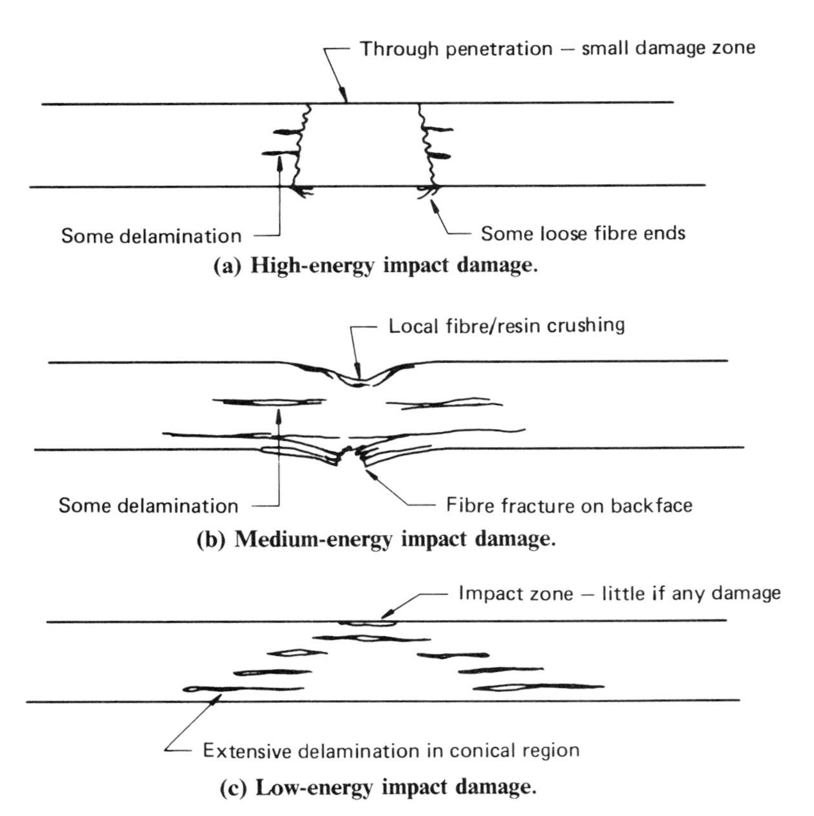

(a) High-energy impact damage.

(b) Medium-energy impact damage.

(c) Low-energy impact damage.

Fig. 10.1 Failure modes in laminated composites resulting from impact at various energy levels.

while boron/epoxy composites exhibited a substantially larger decrease in static strength when compared to metals, the lower density of boron/epoxy provided a weight saving for structures of comparable damage tolerance.

Low-energy impact damage is a major problem in practical fiber-composite structural applications. High and medium levels of impact energy cause surface damage that is relatively easily detected and, therefore, repairs may be undertaken. Low-energy impact can produce damage that is difficult to observe visually (commonly termed "barely visible impact damage" or BVID). This type of damage is of concern because it may occur at quite low-energy levels and is only detectable using nondestructive inspection (NDI) techniques.

The effect of BVID on static and fatigue strengths of graphite/epoxy panels has been studied.[3] It was shown that impact degradation is of more concern in compression loading than tension loading. Specimens tested after impact at a standard energy level were compared with specimens with holes. In tension, the standard energy level impact was equivalent to a 3 mm diameter hole. However, for a similar specimen tested in compression, it was equivalent to a 25 mm diameter hole. Under reversed cycle fatigue ($R = -1.0$), the fatigue performance obtained was lower than for specimens with holes.

Studies[4] have been made of the energy levels at which incipient damage to graphite/epoxy composites occurs. Monolithic specimens of various thicknesses and other specimens with honeycomb cores were subjected to impact under drop weight conditions. Two types of weight drops were used: a blunt spherical indenter and a sharp orthogonal indenter to simulate the impact from the corner of a toolbox. The specimens were instrumented such that the energy-time history was recorded, giving an accurate measure of the energy required to produce the onset of damage. From the results of these tests shown in Fig. 10.2, it can be seen that damage to composites occurs at quite low energy levels and depends upon the thickness of the laminates. As may be expected, a sharp indenter produces damage at a lower energy level than a blunt indenter. (It is important to realize that incipient damage may not necessarily imply damage for which repair is required; this will depend on structural and operational considerations for the component.) Honeycomb panels were shown to be more susceptible to impact damage than monolithic panels of the same thickness. From this and other work,[5] it may be concluded that graphite/epoxy laminates experience damage at quite low energy levels. It was shown that the energy level required to produce damage in graphite/epoxy sandwich panels was half the energy level for damage in similar panels made from S-glass.

To assess the significance of BVID on graphite/epoxy composites in service, low-energy impacts by typical maintenance equipment may be considered. Graphical representations of energy levels for a typical toolbox, spanner (wrench), and screwdriver are shown in Fig. 10.3 plotted against the height of drop. The energy levels for incipient damage for various graphite/epoxy composite laminates are also shown (dotted lines). From these curves, it is apparent that accidental tool drops that would hardly cause concern on metal structures may produce damage on graphite/epoxy structures. For example, a 16 ply monolithic graphite/epoxy laminate

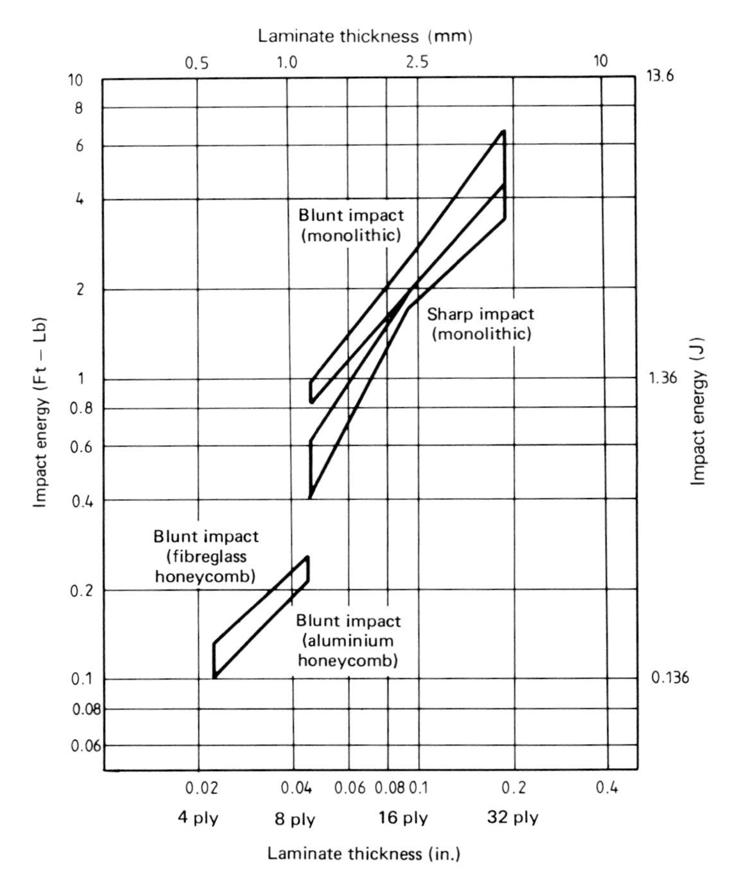

Fig. 10.2 Impact energy for incipient damage to graphite/epoxy laminates. (Taken from Ref. 4.)

experiences damage from a sharp screwdriver dropped 2 m, a sharp-edged spanner dropped 1 m, and a toolbox corner dropped 50 mm. Obviously, upper surfaces of an aircraft are more likely to receive impacts from tool drops than lower surfaces. Since most upper surfaces of an aircraft operate in a compression-dominated load regime and since it has already been shown that impact damage is more significant in composite compression members, consideration is required of maintenance procedures and equipment to provide for the care and protection of composite surfaces during maintenance.

Holes and Slots

Because of the anisotropic nature of composite materials, stress concentrations due to the presence of holes and slots may be substantially higher in composites than for an equivalent metallic structure (see Chap. 7).

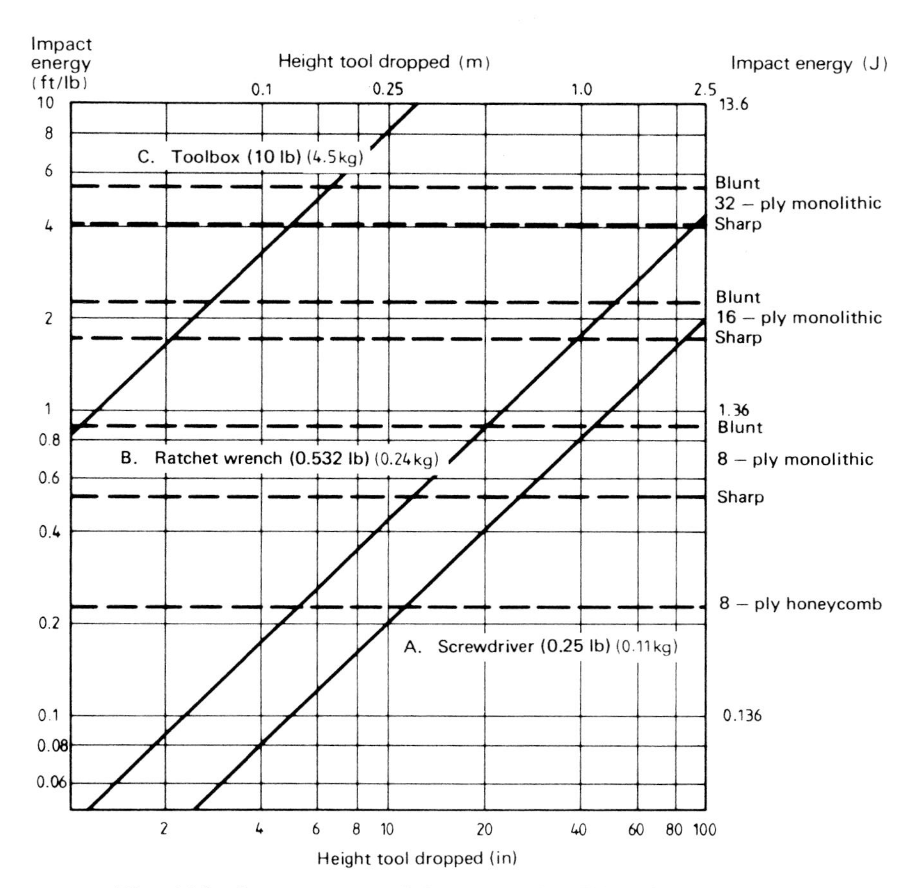

Fig. 10.3 Impact energy of dropped tools. (Taken from Ref. 4.)

Also, fiber composites generally exhibit near-linear elastic behavior to failure. Therefore, the combination of high stress concentrations and the absence of ductile yielding means that composites are relatively intolerant of overloads. The results of experimental evaluations of the effects of holes on fatigue life are discussed in the next subsection and the procedure for the design of composites with holes is given in Sec. 10.3.

Fatigue

In general, tension fatigue in composites is not a problem, with fatigue strengths not greatly below static strength. In compression, larger-diameter fibers such as boron produce substantially better fatigue strengths than smaller-diameter fibers.

The problem of the fatigue of composites with holes has been considered in Ref. 3, which describes experiments involving fatigue regimes of $R = -1.0$

(tension-compression), $R = 0.05$ (tension-tension), and $R = 10.0$ (compression-compression) on graphite/epoxy composites with holes.

Under tension-tension fatigue, specimens achieved 10^6 cycles at maximum loads of up to 90% of ultimate strength, indicating that the tension fatigue of composites with holes is not a major concern. For tests involving compression fatigue, failure lives followed the classical fatigue curves produced by metals. Thus, those materials with holes are susceptible to damage under compression fatigue. Further, similar results were obtained for reversed-cycle fatigue, despite the fact that the stress excursion for this loading is twice the excursion experienced in compression fatigue. Thus, it was concluded that the compressive stress component of a fatigue spectrum is the dominating parameter in determining the fatigue life of composites with holes.

A further point of interest is the effect of fatigue on the residual strength of a composite with holes (see Fig. 10.4). Here the specimens have been subjected to a substantial proportion of their demonstrated fatigue life ($R = -1.0$) prior to tensile testing to failure. From these results, it may be seen that while the specimens were near to failure in fatigue, the tensile strength degradation was minimal. The implications of this with regard to the use of proof testing of structures for recertification after fatigue are significant. An interesting and comprehensive review of compressive fatigue life prediction methodology for composite structures is given in Ref. 6.

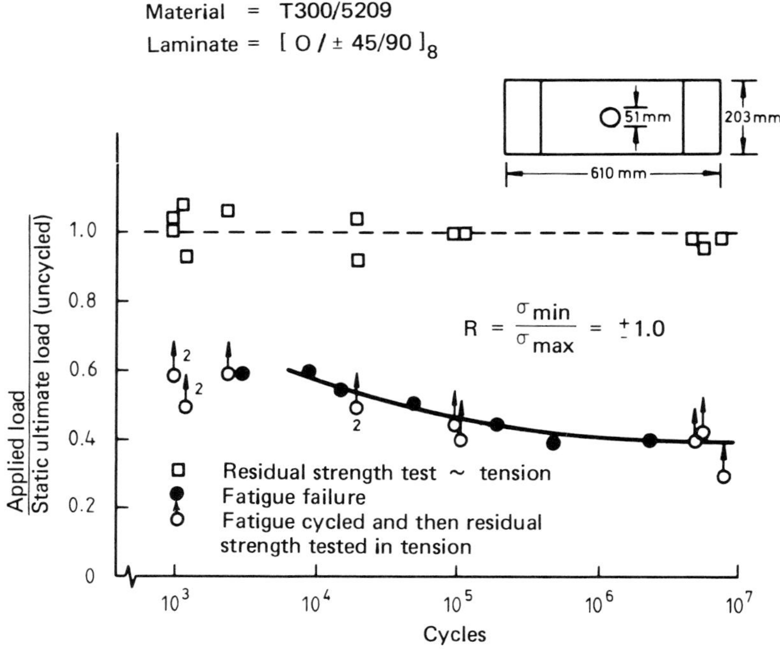

Fig. 10.4 Residual strength of fatigue-cycled panels. (Taken from Ref. 3.)

10.3 DAMAGE-TOLERANT DESIGN

Laminates with Holes

Damage-tolerant design for composite materials takes into consideration the types of "defects" that may, by design, be present in the structure, such as holes, cutouts, and those defects possibly occurring inadvertently during manufacture or service such as delaminations or cracks.

In the case of holes in composite laminates, several different approaches have been successfully developed. These methods are the progressive ply failure approach (e.g., Ref. 7 describes the use of either the Tsai-Hill or the maximum stress failure criterion for individual plies), the average stress failure criterion, and the point stress failure criterion.[8] Of these methods, the latter two seem by far the simplest to use and they will now be described.

Consider a hole of radius R in an infinite orthotropic sheet (Fig. 10.5). If a uniform stress σ is applied parallel to the y axis at infinity, then, as shown in Ref. 8, the normal stress σ_y along the x axis in front of the hole can be approximated by

$$\sigma_y(x,0) = \frac{\sigma}{2}\left\{2 + \left(\frac{R}{x}\right)^2 + 3\left(\frac{R}{x}\right)^4 - (K_T - 3)\left[5\left(\frac{R}{x}\right)^6 - 7\left(\frac{R}{x}\right)^8\right]\right\} \tag{10.1}$$

where K_T is the orthotropic stress concentration factor, which for an infinite sheet is given by

$$K_T = 1 + \left\{2\left[(E_x/E_y)^{\frac{1}{2}} - \nu_{xy}\right] + E_x/G_{xy}\right\}^{\frac{1}{2}} \tag{10.2}$$

The average stress failure criterion[8] then assumes that failure occurs when the average value of σ_y over some fixed length a_0 ahead of the hole first reaches the unnotched tensile strength of the material. That is, when

$$\frac{1}{a_0}\int_R^{R+a_0} \sigma_y(x,0)\,\mathrm{d}x = \sigma_0 \tag{10.3}$$

Using this criterion with Eq. (10.1) gives the ratio of the notched to unnotched strength as

$$\frac{\sigma_N}{\sigma_0} = \frac{2(1-\phi)}{2 - \phi^2 - \phi^4 + (K_T - 3)(\phi^6 - \phi^8)} \tag{10.4}$$

where

$$\phi = R/(R + a_0) \tag{10.5}$$

In practice the quantity a_0 is determined experimentally from strength reduction data.

The point stress criterion assumes that failure occurs when the stress σ_y at some fixed distance d_0 ahead of the hole first reaches the unnotched tensile stress,

$$\sigma_y(x,0)|_{x=R+d_0} = \sigma_0 \tag{10.6}$$

It was shown in Ref. 8 that the point stress and average stress failure criteria are related and that

$$a_0 = 4d_0 \tag{10.7}$$

The accuracy of these methods, in particular the average stress method, can be seen in Fig. 10.6, where a_0 was taken as 3.8 mm. The solid lines represent predicted strength using the average stress criterion, while the dotted lines are the predicted strengths from the point stress method.

With the use of modern structural analyses programs, it is relatively easy to apply these failure criteria to more complex problems. For example, recent tests[9] were carried out on various 16-ply graphite/epoxy laminates (AS/3501-5) with holes. The laminates were: $(0/\pm 45_2/0/\pm 45)_s$, $(0_2/\pm 45/0_2/90/0)_s$, and $(0/\pm 45/90)_{2s}$. The results are shown in Table 10.1 and are compared to predicted values using the average stress method with $a_0 = 2.3$ mm.

As can be seen from the examples given, the average stress criterion provides accurate estimates of the strength reduction due to the presence of holes. This method is widely used in the aerospace industry[10] and has been applied to biaxial stress problems,[11] to the estimation of strength reduction due to battle damage,[12] and to problems in which the stress is compressive.[13]

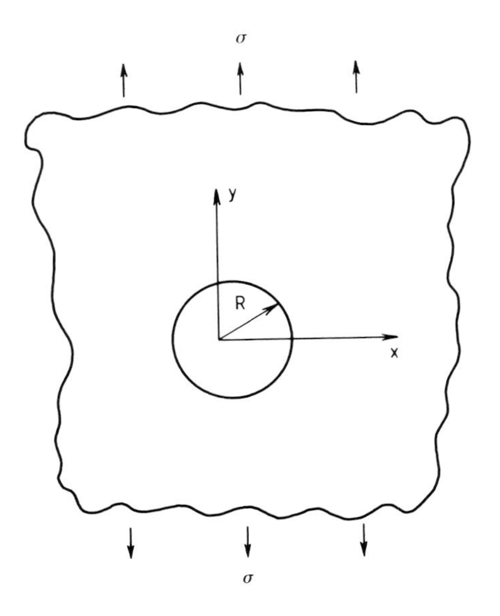

Fig. 10.5 Reference axes for a hole in an orthothropic panel under uniform tension.

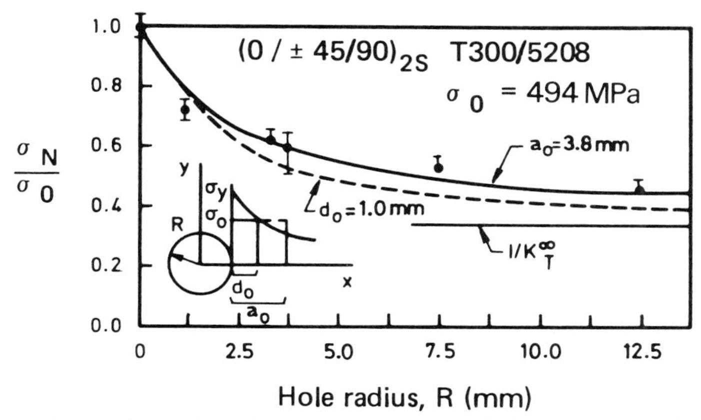

Fig. 10.6 Comparison of predicted and experimental failure stresses for circular holes in $(0/\pm 45/90)_{2s}$ T300/5208.

Table 10.1 Static Strength Predictions[9]

Hole Size and Placement	Laminate No.	% of Unnotched Strength Test	Avg. Stress
2–4.8 mm diameter countersunk, cf. Fig. 10.7a	1	58.9	53.6
	2	48.1	51.4
	3	51.8	53.2
2–4.8 mm diameter countersunk, cf. Fig. 10.7b	3	48.7	45.9
2–6.4 mm diameter countersunk, cf. Fig. 10.7a	3	53.1	50.3
1–6.4 mm diameter noncountersunk	2	54.0	52.6

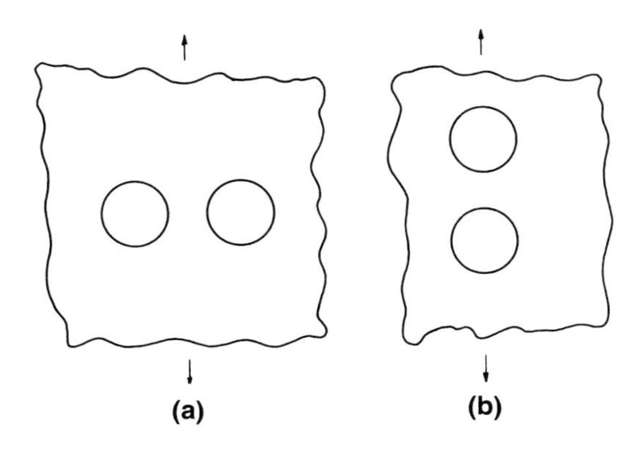

Fig. 10.7 Location of holes with reference to applied load.

Laminates with Cracks

For the problem of through cracks in a fiber composite panel (see Fig. 10.8), there have been a variety of approaches yielding accurate results for the reduction in strength due to the presence of a crack. Each is based on fracture mechanics. The most widely used methods are: the strain energy hypothesis,[14] the compliance approach,[15] and the average and point stress criteria.[8] The first two of these will be discussed only briefly here, with most of the attention being devoted to the third method.

In the strain energy density hypothesis, the strain energy density factor S is defined as

$$S = r_0 \frac{\mathrm{d}W}{\mathrm{d}V} \tag{10.8}$$

where $\mathrm{d}W/\mathrm{d}V$ is the strain energy density and r_0 the radius of a core region surrounding the crack tip. Here,

$$\frac{\mathrm{d}W}{\mathrm{d}V} = \frac{1}{2}\sigma_{ij}\varepsilon_{ij} \tag{10.9}$$

with the usual tensor summation convention.

The assumptions made in the strain energy density approach are:

(1) Crack growth is directed along a line from the center of the spherical core (crack tip) to the point on the surface $r = r_0$ with the minimum strain energy density factor $S_{\min}$.

(2) Growth along this direction begins when S reaches the maximum critical value S_c tolerated by the material.

This approach has been useful in the prediction of failure of bonded joints[14] and unidirectional composites,[16] as well as for angle ply composites.[14] It has also been used for fatigue crack growth predictions.[17]

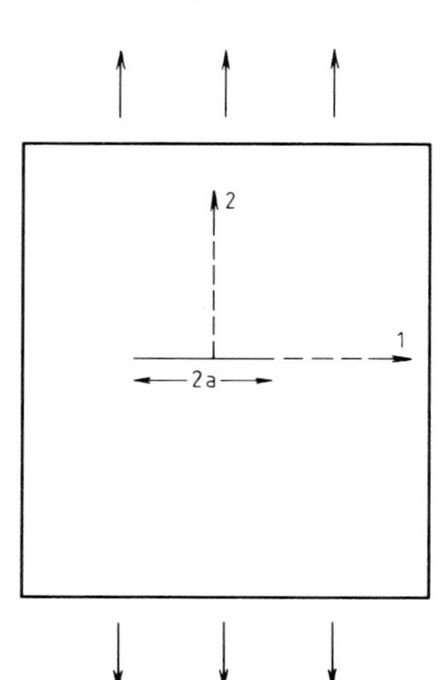

Fig. 10.8 Through-crack in an orthotropic panel.

In the compliance method, the strain energy release rate G is calculated from

$$G = \frac{P^2}{2B}\frac{\mathrm{d}C}{\mathrm{d}a} \tag{10.10}$$

where P is the applied load, $C = \delta/P$, a the crack half-length, B the thickness, and δ the deflection measured between the loading points. Failure occurs when G exceeds a critical value G_c, which is a material constant. For metallic structures, there is a simple relationship between G and the modes I and II stress intensity factors K_{I} and K_{II}. By considering a composite to be a linear elastic orthotropic material with the crack on one plane of symmetry, G is related to K_{I} and K_{II} by

$$G = K_{\mathrm{I}}^2\left(\frac{a_{11}/a_{22}}{2}\right)^{\frac{1}{2}}\left[\left(\frac{a_{22}}{a_{11}}\right)^{\frac{1}{2}} + \frac{2a_{12}+a_{66}}{2a_{11}}\right]^{\frac{1}{2}} + K_{\mathrm{II}}^2 a_{11}\left[\left(\frac{a_{22}}{a_{11}}\right)^{\frac{1}{2}} + \frac{2a_{12}+a_{66}}{2a_{11}}\right]^{\frac{1}{2}}\left(\frac{1}{2}\right)^{\frac{1}{2}} \tag{10.11}$$

where a_{ij} are the elements of the compliance tensor with the 1 subscript in the direction of the crack. This procedure has been widely used.[18]

In the average stress method, the analysis proceeds as follows. For cracks in orthotropic laminates, the stress along the net section beyond the crack tip is given by

$$\sigma_y(x,0) = Y\sigma x/(x^2 - a^2)^{\frac{1}{2}} \tag{10.12}$$

where σ is the applied stress, a the crack half-length, and Y the finite width correction factor. Making use of this, the average stress failure criterion states that failure occurs when

$$\frac{1}{a_0}\int_a^{a+a_0} \sigma_y\,\mathrm{d}x = \sigma_0 \tag{10.13}$$

which, after using Eq. (10.12), gives

$$Y\sigma_N/\sigma_0 = [(1-\xi)/(1+\xi)]^{\frac{1}{2}} \tag{10.14}$$

where

$$\xi = a/(a + a_0) \tag{10.15}$$

and a_0, as previously, is the length of the damage zone in front of the crack. If the point stress criterion is used, this gives

$$\sigma_N/\sigma_0 = (1 - \xi_1^2)^{\frac{1}{2}} \tag{10.16}$$

where

$$\xi_1 = a/(a + d_0)$$

As previously discussed, the two criteria are compatible if $d_0 = a_0/4$. These criteria are relatively simple and are widely used. Table 10.2 shows the experimental tensile strengths and the fracture toughness for various laminate constructions as well as the damage zone sizes d_0. Table 10.3 shows similar results, but for cracks oriented at an angle to the load direction. From these and other works,[19,20] it may be seen that the average and point stress methods can accurately predict the reduction in strength due to the presence of cracks. Other interesting works in this field are given in Refs. 21 and 22. Also, see Fig. 10.9.

A further extension of the average stress criterion may be considered. For a notched orthotropic laminate, the fracture toughness K_{1c} is given by

$$K_{1c} = \sigma_N (\pi a)^{\frac{1}{2}} \tag{10.17}$$

where σ_N is the stress applied to the notched panel at failure and a the crack half-length. This may be expressed in terms of the unnotched strength σ_0 by use of Eq. (10.14),

$$K_{1c} = \sigma_0 [\pi a(1 - \xi)/(1 + \xi)]^{\frac{1}{2}} \tag{10.18}$$

It may be shown that as a becomes large, Eq. (10.18) approaches an asymptote,

$$K_{1c} \rightarrow \sigma_0 (\pi a_0/2)^{\frac{1}{2}} \tag{10.19}$$

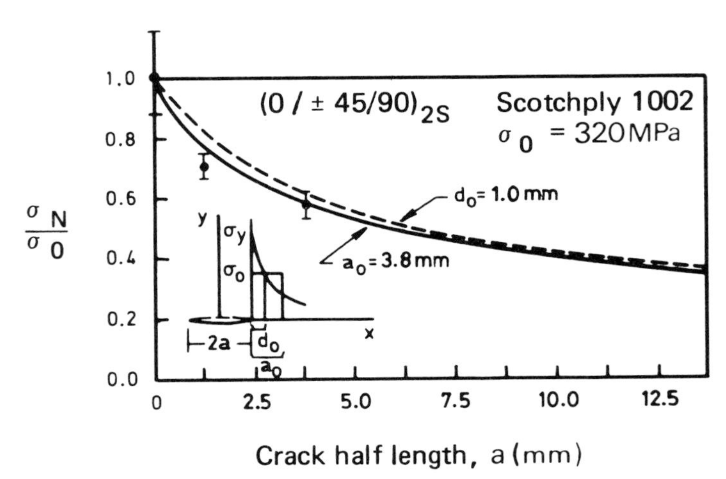

Fig. 10.9 Comparison of predicted and experimental failure stresses for center cracks in $(0/\pm 45/90)_{2s}$ Scotchply 1002. (Taken from Ref. 12.)

As would be expected, K_{1c} is independent of the crack length for large cracks. However, as $a \to 0$, $K_{1c} \to 0$. Since K_{1c} should remain constant and be independent of crack length, it is necessary to introduce a correction factor C_0, to provide a modified critical stress intensity K'_{1c},

$$K'_{1c} = \sigma_N[\pi(a + C_0)]^{\frac{1}{2}} \tag{10.20}$$

This is analogous to the Irwin-type correction factor for plastic zone effects in ductile materials. Since fiber composites do not yield, C_0 may represent a damage zone ahead of the crack in which matrix and fiber cracking occur. Use of Eq. (10.14) again gives

$$K'_{1c} = \sigma_0[\pi(a + C_0)(1 - \xi)/(1 + \xi)]^{\frac{1}{2}} \tag{10.21}$$

Here, as $a \to 0$, $K'_{1c} \to \sigma_0(\pi C_0)^{\frac{1}{2}}$. By taking $C_0 = a_0/2$, a constant value of K'_{1c} is obtained, which is independent of crack length,

$$K'_{1c} = \sigma_0(\pi a_0/2)^{\frac{1}{2}} \tag{10.22}$$

Table 10.2 Tensile Strength and Fracture Toughness[20]

Laminate Construction	Tensile Strength, MPa	Fracture Toughness, $MPa \cdot m^{\frac{1}{2}}$	d_0, mm
T300/5208 graphite/epoxy			
$[0/90]_{4s}$	638	13.0	1.01
$[0/\pm 45]_{4s}$	542	39.8	1.01
$[0/\pm 45/90]_{2s}$	467–494	33.2	—
E-glass/epoxy (Scotchply 1002)			
$[0/90]_{4s}$	424	30.7	1.01
$[0/\pm 45/90]_{2s}$	321	24.3	1.01
Boron/epoxy			
$[0_2/\pm 45]_{2s}$	697	62.0	1.27
$[0_3/\pm 45/90]_{2s}$	691	59.8	1.27
$[0/\pm 45]_{2s}$	608	49.9	1.27
$[0/\pm 45/90]_{2s}$	421	38.7	1.27
AS/3501 graphite/epoxy			
$[0]_{8s}$	1188	90.7	1.93
$[0/90]_{4s}$	500	24.6	0.38
$[0/\pm 60]_{4s}$	452	28.0	0.61
$[0/\pm 45/90]_{2s}$	480	32.8	0.75
$[0/\pm 36/\pm 72]_{2s}$	417	36.2	1.2
$[0/\pm 30/\pm 60/90]_{2s}$	466	27.2	0.45

Table 10.3 Static Tensile Strength Predictions and Test Results for Slant and Normal Cracks in Graphite/Epoxy AS/3501-5 (from Ref. 19)

Laminate Construction	Crack Orientation to Load, deg	% Unnotched Tensile Strength Test	Avg. Stress	d_0, mm
$(0_2/\pm 45/0_2/90/0)_s$	45	53.4	50.5	0.575
$(0/\pm 45/\pm 90)_{2s}$	45	51.4	50.7	0.575
$(0_2/\pm 45/0_2/90/0)_s$	90	44.1	44.2	0.575
$(0/\pm 45/\pm 90)_{2s}$	90	47.2	44.3	0.575

Ballistic Damage

So far, the problems of holes and cracks have been discussed. However, fracture mechanics has also been shown[12, 23, 24] to be applicable to the estimation of the residual strength of ballistically damaged composite panels. The residual strength of damaged tension panels has been derived[12] from the point stress criterion in the form

$$\sigma/\sigma_0 = [(W_s - KW_{ke})/W_s]^{\frac{1}{2}} \tag{10.23}$$

where σ is the residual strength of the damaged panel, σ_0 the unnotched strength, K an experimentally determined constant, W_s the strain energy derived from the stress-strain curve for the unnotched material, and W_{ke} the kinetic energy per unit thickness imparted to the structure. This model is applicable for the range of impact velocities that do not result in complete penetration. Where through penetration occurs, Eq. (10.4) may be used to estimate the reduction in strength.

The parameter K is dependent on the ratio of the panel width to the projectile diameter. However, tests[12] have shown that K rapidly asymptotes to a constant value.

Material Variability

Unnotched condition. For an orthotropic material, the critical stress intensity K'_{1c} is given in Eq. (10.20) as

$$K'_{1c} = \sigma_N[\pi(a + C_0)]^{\frac{1}{2}} \tag{10.24}$$

From this it is evident that, in the unnotched condition ($a = 0$), failure is due to an "inherent flaw" of size $C_0 = a_0/2$. Therefore, the fracture toughness may be affected by factors influencing the inherent flaw size in the unnotched condition. Obviously, one important factor that will influence the inherent flaw size is the quality of the original laminate; a poor-quality laminate (with, say, a high void content) will exhibit a larger inherent flaw size than a good-quality laminate. The reduction in strength due to quality is given by use of Eq. (10.22),

$$\sigma_0'/\sigma_0 = (a_0/a_0')^{\frac{1}{2}} \tag{10.25}$$

where σ_0' is the unnotched strength of the poor-quality laminate and a_0' the characteristic length.

Notched laminates. The problem of the effect of quality of the laminate in the presence of a large defect, such as a crack, needs consideration from two views: the strength reduction of the poor-quality laminate, and the residual strength reduction due to the defect together with the laminate quality, when compared to the unnotched strength of a good-quality laminate. The strength reduction of the poor-quality laminate may be determined by evaluating a_0', the characteristic length for the poor-quality laminate. This is then substituted in Eq. (10.14) to give

$$\sigma_N'/\sigma_0' = [a_0'/(2a + a_0')]^{\frac{1}{2}} \tag{10.26}$$

where σ_N' is the notched strength of the poor-quality laminate. From this, it may be seen that as a_0' increases (laminate quality decreasing), σ_N'/σ_0' increases; thus, the material becomes less sensitive to the presence of the flaw. (While σ_N'/σ_0' increases with a_0', the individual values of σ_N' and σ_0' will decrease.)

The comparison of notched strength of the poor-quality laminate, when compared to an unnotched good-quality laminate is given by

$$\sigma_N'/\sigma_0 = [a_0/(2a + a_0')]^{\frac{1}{2}} \tag{10.27}$$

Here, it is apparent that as laminate quality decreases, the notched strength decreases when compared to an unnotched good-quality laminate.

Delaminations

Delaminations in layered composite materials may occur due to a variety of reasons, such as low energy impact or manufacturing defects. The presence of delaminations is of major concern in the vicinity of bonded joints and in compressively loaded components where the delaminations may grow under fatigue loading by out-of-plane distortion.

An early study into the growth of delaminations[25] arose from the B-1 composite development program. The effects of delamination size, temperature, and moisture on AS/3501-5A graphite/epoxy panels was studied. From these tests, it is clear that compressive strength degrades with the increasing size of the delamination and with increasing temperature and moisture content. For example, one laminate $(90/0/\pm 45)$, with a 12.5 mm delamination in a 38.1 mm wide specimen, showed a 20% compressive strength reduction in a room environment. A similar specimen tested at 132°C and 1.3% moisture content with the same size delamination gave a 50% compressive strength reduction. Recent tests[26] on the fatigue growth of a 50.8 mm diameter delamination (due to impact damage in a wing box) showed that the effect of impact damage on a full-scale structure can be estimated from specimen tests.

To date there have been few analytical studies into delamination growth. A detailed study[27] used the NASTRAN finite element program to determine interlaminar stresses, which were subsequently used to predict accurately delamination growth rates. A similar study[28] into edge delaminations was undertaken, using the strain energy release rate G to predict delamination growth rates. In this work, it appears that the energy release rate remains relatively constant during delamination growth. A detailed three-dimensional analysis[29] of stresses developed during impact has also been undertaken. Sandwich panels with fiber composite faces and a thick laminate were considered. For the laminate, impact develops large trans-

Table 10.4 Compressive Strength of Various Impact-Damaged Laminates[a]

Lay-up[b]	Skin Thickness, mm	Delamination Area, mm^2	Compressive Failure Strain, μ in./in.
AS/3501-6 $(\pm 45/0_2/\pm 45/0_2/\pm 45/0/90)_{2s}$	6.91	1290	3780
AS/3501-6 $(+45/90/-45/+22.5/-67.5/-22.5/+67.6/\pm 45/+67.5/+22.5/-67.5/-22.5/\pm 67.5/\pm 22.5/0_2/\pm 22.5)_s$	6.45	1613	4630
AS/3501-6 $(0/\pm 45/0_2/\mp 45/0)$ skin on HRP-3/16-5.5 honeycomb core	1.07	1161	4270
AS/3501-6 $(\pm 45/0^*/\pm 45/0_2^*/90^*/0_2^*/\pm 45^*/0_2^*/\pm 45^*/0^*/+45^*/0^*/\mp 45^*/0^*/-45^*/0^*/\pm 45^*)_s$	12.7	2389	4090
AS/3501-6 $(90^*/-45^*/\mp 67.5/\mp 22.5^*/0/90^*/+22.5^*/0_2/\pm 45^*/0^*/-22.5^*/+45^*/+67.5^*/+22.5^*/-67.5^*/-45^*/+22.5_2^*/0)_s$	11.1	7097	5460
T300/5208 $(\pm 45/0_2/\pm 45/0_2/\pm 45/0/90)_{2s}$	6.71	1419	4000
T300/5208 $(\pm 45/0/90/\mp 45/0/90)_{3s}$	6.71	1226	4000

[a] From Ref. 26.
[b] Specimens were impact tested and then loaded statistically to failure.
*Double-ply material. (All other materials cited are single ply.)

verse shear stresses in the damage region and it appears that the subsurface transverse shear stresses are critical. In the case of the sandwich panel, the subsurface 0° plies would be critical.

Finite element analysis has also been used to investigate the static compressive failure of delaminations.[30] In this study, a detailed three-dimensional finite element analysis of a delaminated graphite/epoxy (namely, T300/5208) beam is performed. The orientation of the plies is $(\pm 45/0_2)_{2s}$ and delaminations 12.7, 25.4, and 38.1 mm long are considered. The location of the delamination is varied. The maximum strain energy density S at each point along the delamination is computed as is the energy release rate G. The far-field failure strain is then estimated on the basis that at failure $G_c = 7.5\ \mathrm{N \cdot m^{-1}}$ and/or $S_c = 9.64\ \mathrm{N \cdot m^{-1}}$, as given in Refs. 31 and 14, respectively. Both failure criteria give a far-field failure strain of approximately 4.5×10^{-3} for a delamination 38.1 mm long between the near-surface $\pm 45°$ plies and the 0_2 plies. The failure strain for such a delamination is in good agreement with that given in Ref. 26 and again confirms the conclusion given there that estimates of the critical flaw size due to delaminations or impact damage may be obtained from specimen tests. Indeed, as can be seen from Table 10.4, static failure has been found to occur for a variety of laminates and fiber orientations at approximately 4000 microstrain. Furthermore it has now been clearly shown in Refs. 25, 30, and 32 that delamination growth under compressive loading is entirely due to the unsymmetrical nature of the plies above and below the delamination. The compressive forces acting on these plies produce out-of-plane bending and it is this bending that drives the delamination.

Approximate analytical methods other than the finite element method have also been used.[33-35] These methods are capable of differentiating between laminates whose stacking sequence is susceptible to delamination growth and those laminate stacking sequences more resistant to delamination.

The cause of the final failure of the bulk material is still uncertain, although a number of hypotheses have been proposed. It is most probable that failure is due to a combination of out-of-plane bending due to loss of symmetry and a reduction of the net cross section.

10.4 DAMAGE TOLERANCE IMPROVEMENT

Various papers have shown methods for damage tolerance enhancement of composite materials. It has been shown[36] that the surface treatment of graphite fibers to enhance bond strength produces a decrease in izod impact fracture energy. It is believed that this is due to an alteration in the failure mode, which for untreated fibers involves more delaminations and therefore greater energy absorption. This work also showed that, by use of a plasticizer in the resin system, a higher strain-to-failure matrix was produced, which resulted in an increase in the fracture energy.

The effects of the laminate stacking sequence on fracture strength have been considered,[3] using specimens with the same number of oriented plies, but with differing stacking sequences. The specimens were also slotted. While design procedures would suggest that the laminates would ostensibly have the same unnotched strength, the fracture strengths of the specimens varied by up to 30%. Stronger panels were those with the outer layers at 0° or 90° to the load direction. Specimens with ±45° fibers on the outer faces produced lower fracture strengths. Also, bunching of 0° layers together was shown to produce higher notched strengths, since extensive delamination was possible between the 0° layers, which absorbs more energy.

A major cause of impact related problems in high-performance composites is their low strain to failure. Honeycomb face sheets of graphite epoxy have a strain to failure about one-third of that for S-glass face sheets. Various works have evaluated the concept of hybridization, i.e., mixing the fiber types, thus utilizing the advantageous properties of one material to overcome deficiencies of another. Glass fiber face sheets on a graphite core have been compared to an all graphite structure.[36] The higher strain to failure of glass fibers led to substantial increases in threshold energies for impact damage, with the energy level required to produce delamination being increased by a factor of four.

The more recent availability of aramid fibers has added impetus to studies in hybrid composite systems. Aramid fibers (generally known under their trade name Kevlar) are relatively inexpensive and exhibit a high strain to failure, but also have a substantially higher modulus of elasticity than glass fibers. Therefore, they exhibit a high tensile strength without the stiffness penalty of glass fiber composites. Most work using this fiber has concentrated on the aramid/graphite/epoxy hybrid, since the aramid material provides an increase in impact resistance and a reduced material cost, while the graphite fibers provide compressive strength, which the aramid material lacks. Some studies[37] have shown that aramid/graphite hybridization leads to a doubling of residual strength after impact compared to all-graphite specimens. For unidirectional lay-ups, the same comparison showed an improvement by a factor of three.

A further concept for improvement of notch sensitivity is that of specific hybridization. The incorporation of strips with a high strain to failure in a structure (softening strips) acts to produce a zone where fracture of the structure is arrested by the high displacement capability of the strips. Similarly, the presence of high-modulus strips (hardening strips) provides a restraint to opening of a crack, thus assisting arrest of the crack.

References

[1]Avery, J. G. and Porter, T. R., "Comparison of the Ballistic Response of Metals and Composites for Military Aircraft Applications," ASTM STP 568, 1975, pp. 3–29.

[2]Suarez, J. A. and Whiteside, J. B., "Comparison of Residual Strength of Composite and Metal Structures After Ballistic Damage," ASTM STP 568, 1975, pp. 72–91.

[3]Walter, R. W., Johnson, R. W., June, R. R., and McCarthy, J. E., "Designing for Integrity in Long-Life Composite Aircraft Structures," ASTM STP 636, 1977, pp. 228–247.

[4]Naval Air Engineering Center, NAEC Rept. 92–136.

[5]Oplinger, D. W. and Slepetz, J. M., "Impact Damage Tolerance of Graphite Epoxy Sandwich Panels," ASTM STP 568, 1975, pp. 30–48.

[6]Saff, C. R., "Compression Fatigue Life Prediction Methodology for Composite Structures-Literature Survey," NADC Rept. 78203-60, June 1980.

[7]Soni Som, R., "Failure Analysis of Composite Laminates with a Fastener Hole," AFWAL Rept. TR-80-4010, 1980.

[8]Nuismer, R. J. and Whitney, J. M., "Uniaxial Failure of Composite Laminates Containing Stress Concentrations," ASTM STP 593, 1975, pp. 117–142.

[9]Nuismer, R. J. and Labor, J. D., "Applications of the Average Stress Failure Criterion, Part 1: Tension; *Journal of Composite Materials*, Vol. 12, 1978, p. 238.

[10]Pimm, J. H., "Experimental Investigation of Composite Wing Failure," AIAA Paper 78-509, 1978.

[11]Daniel, I. M., "Behaviour of Graphite Epoxy Plates with Holes under Biaxial Loading," *Experimental Mechanics*, Vol. 20, Jan. 1980, pp. 1–8.

[12]Husman, G. E., Whitney, J. M., and Halpin, J. C., "Residual Strength Characterization of Laminated Composites Subjected to Impact Loading," ASTM STP 568, 1975, pp. 92–113.

[13]Nuismer, R. J. and Labor, J. D., "Application of the Average Stress Failure Criterion: Part 2—Compression," *Journal of Composite Materials*, Vol. 13, Jan. 1979, pp. 49-60.

[14]Sih, G. C., "Mechanics of Fracture: Linear Response," *Numerical Methods in Fracture Mechanics*, *Proceedings of First International Conference*, University College Swansea, UK, edited by A. R. Luxmore and D. R. J. Owen, 1978, pp. 155–192.

[15]Slepetz, J. M. and Carlson, L., "Fracture of Composite Compact Tension Specimens," ASTM STP 593, 1975, pp. 143–162.

[16]Sih, G. C. and Chen, E. P., "Fracture Analysis of Unidirectional Composites," *Journal of Composite Materials*, Vol. 7, 1973, pp. 230–244.

[17]Badaliance, R. and Drill, H. D., "Compression Fatigue Life Prediction Methodology for Composite Structures, Vol. II: Technical Proposal, McDonnell Aircraft Co., Rept. MDC-A573, Feb. 1979.

[18]Barnby, J. T. and Spencer, B., "Crack Propagation and Compliance Calibration in Fibre-Reinforced Polymers, *Journal of Materials Science*, Vol. 11, 1976. pp. 78–82.

[19]Morris, D. M. and Hahn, H. T., "Mixed Mode Fracture of Graphite/Epoxy Composites: Fracture Strength," *Journal of Composite Materials*, Vol. 11, 1977, p. 124.

[20]Caprino, G., Halpin, J. C., and Nicolais, L., "Fracture Mechanics in Composite Materials," *Composites*, Vol. 10, Oct. 1979, pp. 223-227.

[21]Dorey, G., "Damage Tolerance in Advanced Composite Materials," Royal Aircraft Establishment, Tech. Rept. TR 77172, Nov. 1977.

[22]Kanninen, M. F., Rybicki, E. F., and Brinson, H. F., "A Critical Look at Current Applications of Fracture Mechanics to the Failure of Fibre-Reinforced Composites," *Composites*, Vol. 8, Jan. 1977, pp. 17–22.

[23]Olster, E. F. and Roy, P. A., "Tolerance of Advanced Composites to Ballistic Damage," ASTM STP 546, 1974, pp. 583–603.

[24]Dorey, G., Sidey, R., and Hutching, J., "Impact Properties of Carbon Fibre/Kevlar Reinforced Plastic Hybrid Composites," Royal Aircraft Establishment, Tech. Rept. 76057, 1976.

[25]Konishi, D. Y. and Johnston, W. R., "Fatigue Effects on Delaminations and Strength Degradation in Graphite/Epoxy Laminates," ASTM STP 674, 1979, p. 597.

[26]Gause, L. W., Rosenfeld, M. S., and Vining, R. E. Jr., "Effect of Impact Damage on the XFV-12A Composite Wing Box", *Proceedings of 25th National SAMPE Symposium*, Vol. 25, 1980, pp. 679–690.

[27]Ratwani, M. M. and Kan, H. P., "Compression Fatigue Analysis of Fibre Composites," NADC Rept. 78049-60, Sept. 1979.

[28]Rybicki, E. F., Schmueser, D. W., and Fox, J., "An Energy Release Rate Approach for Stable Crack Growth in Free-Edge Delamination Problem," *Journal of Composite Materials*, Vol. 11, 1977, p. 470.

[29]Stanton, E. L. and Crain, L. M., "Interlaminar Stress Gradients and Impact Damage," *Fibrous Composites in Structural Design*, edited by E. M. Lenoe et al., Plenum Press, New York, 1980, pp. 423–440.

[30]Jones, R. and Callinan, R. J., "Analysis of Compression Failures in Fibre Composites," *Proceedings of ICCM4*, Japan Society for Composite Materials, Tokyo, Oct. 1982, pp. 447–454.

[31]Ramkumar, R. L., Kulknarni, S. V., and Pipes, R. B., "Definition and Modelling of Critical Flaws in Graphite Fibre Reinforced Epoxy Resin Matrix Composite Materials," NADC-76228-30, 1976.

[32]Verette, R. M. and Demuts, E., "Effects of Manufacturing and In-service Defects on Composite Materials," *Proceedings of Army Symposium on Solid Mechanics, 1976—Composite Materials: The Influence of Mechanics of Failure on Design*, AMMRC-MS-76-2, 1976, pp. 123–137.

[33]Rodini, B. T. and Eisenmann, J. R., "An Analytical and Experimental Investigation of Edge Delamination in Composite Laminates," *Fibrous Composites in Structural Design*, edited by E. M. Lenoe et al., Plenum Press, New York, 1980, pp. 441–458.

[34]Pagano, N. J. and Pipes, R. B., "Some Observations on the Interlaminar Strength of Composite Laminates," *International Journal of Mechanical Sciences*, Vol. 15, 1973, p. 679.

[35]Pipes, R. B., "Interlaminar Strength of Laminated Polymeric Matrix Composites," AFML-TR-7682, 1976.

[36]Bradshaw, J., Dorey, G., and Sidey, R., "Impact Resistance of Carbon Fibre Reinforced Plastics," Royal Aircraft Establishment, Tech. Rept. 72240, 1973.

[37]Dorey, G., "Improved Impact Tolerance in Composite Structures," *Symposium on the Design and Use of Kevlar[R] in Aircraft*, Geneva, Oct. 1980, Chap. 9.

11. NDI OF FIBER-REINFORCED COMPOSITE MATERIALS*

11.1 INTRODUCTION

Nondestructive inspection (NDI) of fiber-reinforced (FR) composite materials can be expected to differ from that of metallic materials because composites themselves differ markedly from metals and their alloys. FR composites are inhomogeneous and markedly anisotropic, possess a low thermal conductivity along with a high acoustic attenuation, and are generally poor conductors of electricity. High-performance structures are conventionally made from metallic material that is relatively free of unwanted defects; in-service failures tend to originate from crack initiation at identifiable defects and occur after crack propagation. Hence, NDI procedures can be based on the detection/location of growing cracks, the importance of which can be determined using fracture mechanics. No similar predominant failure process has yet been identified for composite material, no procedure similar to fracture mechanics has been developed, and many of the NDI needs are as yet not clearly defined.

11.2 THE SEARCH FOR DEFECTS

For composite materials, much of the NDI is conducted either during, or immediately after, the manufacture of the component and consists of looking for delaminations, debonds, etc. Subsequently, damage to composites can accrue from impact, environmental effects, or static or cyclic load application, all of which have different effects on the defects already present or may introduce different types of defects altogether.

Table 11.1 lists the defects (and related factors) that are likely to affect the strength of composite material. A description of the NDI techniques applicable to the detection of these defects is given in Table 11.2 along with a summary of principles of operation, capabilities, and disadvantages of each technique. All of the techniques listed in Table 11.2 will be discussed further, particularly the most commonly used ones—x-radiography and ultrasonic C-scan. The remainder of those shown are still undergoing development or are not as well established, either as NDI techniques or for specific application to composite materials.

---

*This chapter is a condensation of the comprehensive review given in Ref. 1.

Table 11.1 Factors Affecting Strength of Fiber-Reinforced Composites

Factors Related to Manufacturing Processes
Variation of fiber/matrix ratio; condition of fiber/matrix interface; variation in matrix properties
Thermal decomposition of matrix; local undercure of matrix
Lay-up defect: error in ply or fiber orientation, missed plies, gaps between plies, or excessive ply overlap
Dimensional error; inclusion of a foreign object; surface fault/scratch
Delamination; translaminar cracking
Factors Arising from Service Use
Degradation of matrix by environment, especially the effect of moisture absorption
Creep
Damage zone around stress concentration; impact damage
Fracture or buckling of fiber; loss of fiber/matrix bond; delamination; matrix crazing; cracks arising from local overload or fatigue processes

X-Radiography

X rays are differentially absorbed in passing through a material according to the atomic number of the elements present. For the organic composite materials, difficulty may be expected in obtaining x-ray images of adequate contrast. Typically, low-energy x rays (a few tens of kiloelectron volts) are used with a beryllium window for composites testing. Currents are held to a few milliamperes and exposures to about 1 min. Specimen-to-generator distances vary greatly, usually 0.3–3 m.

There appear to be contradictory findings about what types of defects can be detected by x-radiography. Harris[2] claims that radiography can be used to identify voids, that fiber/resin debonding or cracks resulting from thermal contraction cannot be distinguished, and that interlaminar cracks cannot be identified. Prakash[3] suggests that thermal cracks are readily detected, as are foreign objects and inclusions, but that interlaminar debonds and fiber debonds are generally hard to detect. Salkind[4] was unable to find void or crack indications in his tests but Nevadunsky et al.[5] were able to detect large voids, cracks, and porosity in adhesives.

Determination of resin content in a composite should be feasible using x-radiography. However, consideration must be given to the relative absorption coefficients of the particular fiber and matrix in the composite under investigation. Martin[6] determined mass absorption coefficients for a graphite/epoxy composite, both theoretically and experimentally (using attenuation measurements), in an attempt to obtain resin contents. However, neither this method nor an attempt to relate measured film density to resin content worked very well. Forli and Torp[7] studied a glass/epoxy composite in which the absorption coefficient was 20 times higher in the glass than in unpigmented polyester. They were able to determine the total glass content from film density measurements and also to observe the reinforcement type and orientation.

Enhancement of some forms of cracks for radiographic testing is possible, but the effects on life have yet to be established. Hagemaier and Fassbender[8] tested specimens containing edge delaminations with diiodobutane (DDB) or s-tetrabromethane (TBE). Radiographs taken an hour later were enhanced and the chemicals evaporated completely after 2–3 days. However, they noted that DDB was an irritant, while TBE was a severe poison and a potent mutagent which should not be used in the future.

Mention is often made of penetration of suitable materials into voids, but the process must surely be very slow unless there is coalescence of microvoids forming a channel to a free surface of the material.

The contrast available from conventional radiographic techniques is usually insufficient to permit resolution of individual fibers. However, Roderick and Whitcomb[9] placed the x-ray source close to a boron/epoxy laminate and a high-resolution glass plate and were able to discern breaks in the tungsten cores ($\sim 3\ \mu m$ diameter) of the boron fibers resulting from fatigue testing. Crane et al.[10] proposed that boron fibers be added to the edge of each composite tape of graphite/epoxy so that the fiber alignment could be assessed. Crane[11] also proposed using the fringe patterns, which appear in radiographs due to misaligned fibers, to measure misalignment.

Ultrasonic C-Scan

The principal established ultrasonic NDI technique for composite materials is the C-scan in which a plane in the object is scanned. An ultrasonic pulse is propagated through the specimen, reflected from a back surface or a defect, and received (usually) by the transmitting sensor. The ultrasonic pulse is scattered or reflected from any defect that differs greatly in acoustic impedance from the surrounding material. The readily available information in the echo signal is the amplitude and the time for the signal to travel from the transmitter to the receiver. The latter may be measured by means of an electronic gate. The width of the gate may be set to select only a thin plane within the specimen for examination (a similar effect can be obtained by means of focusing transducers) or to average an acoustic parameter through the thickness of the specimen. C-scan testing is usually done in a water tank, although bubbler techniques are available for large structures. A permanent record is available from a Mufax or similar recorder fitted with line-intensification capabilities for amplitude recording. Complications arise when the specimen surfaces are either curved or nonparallel, although Liber et al.[12] claim that these are largely overcome by using front-surface triggering of the gate.

Interrogation of a specimen by ultrasonic pulses yields information relating to material acoustic properties and specimen dimensions. Dimensions can be obtained from the signals reflected from the front and back faces of the specimen; information can also be obtained about nonplanar

Table 11.2 NDI Techniques to Detect

Technique	Operating Principles
X-radiography	A beam of x rays is differentially absorbed as it passes through a specimen; rates of absorption are greatest for metals and vary roughly according to atomic number.
Ultrasonic C-scan	An ultrasonic pulse, propagating through an object, is scattered or reflected from any interface that separates regions of differing acoustic impedance. The pulse is scattered from a defect or reflected from the front and back surfaces of an object. Amplitude and transit time of reflections yield defect information.
Neutron radiography	A beam of low energy (or thermal) neutrons is differentially absorbed as it passes through a specimen; rates of absorption are greatest for hydrogen and the organic materials used in composites.
Optical holography	Coherent light from a laser is used to form a hologram of the specimen surface; a speckle pattern can be formed to give surface information or the hologram of a changed surface can be viewed through the hologram of the original surface (live fringes) or two holograms can be compared (frozen fringes).
Acoustic holography	Acoustic holography is analogous to optical holography, but in this case the acoustic wave scattered or reflected by an object interferes with an acoustic reference signal. Thus, both amplitude and phase information is recorded.
Thermography	Temperature changes appearing at the specimen surface are measured using sensitive infrared equipment; temperature changes arise from distortion of an injected heat field by defects or as part of fatigue process or from rubbing of surfaces produced by an applied vibration.

efects in Composite Materials

Capabilities	Disadvantages
Noncontact, but remote capability is dependent on factors such as focusing, sensitivity, etc.; internal defects can be located and occasionally identified. Maximum sensitivity is achieved when the defect is parallel to the x-ray beam.	Crack detection sensitivity is strongly dependent on crack orientation; contrast between epoxy and carbon rarely permits examination of individual fibers. Distortion that arises at edges of collimated beam of x rays limits field of view. Discrimination between different types of defect may be difficult.
Delaminations in a plane normal to the ultrasonic vave beam can be detected. Cracks can be letected, depending upon their orientation. Large hree-dimensional voids are easily detected.	Ultrasound is introduced to specimen via water bath or water jet; water may have a deleterious effect on material particularly if edge delaminations are present. Dependence of sensitivity on orientation may be a problem. Technique is unable to distinguish between delaminations and voids without multiple measurements.
Organic materials are readily examined in close proximity to metals; corrosion products are also emphasized.	Use of nuclear reactor requires specimens to be taken to a reactor site. Hence, a portable source is required, together with adequate safety precautions. Californium has been suggested, but is expensive and has a short half-life.
Surface deformations or surface strains can be letermined using a device which is remote from specimen [noncontact]; sensitivities are dependent on mechanical stability of test platform and specimen.	Remains essentially a laboratory technique because of problems with any ambient vibration.
Available as commercial equipment. Detection capabilities are as for C-scan, but identification of defects is much better because a three-dimensional picture can be obtained.	Surface defects may cause problems of interpretation.
Detects large defects using "at distance" equipment that frequently can be operated without affecting plant or structure operation.	Thermal techniques are notoriously troubled by changes in ambient conditions; most commonly used equipment is rather expensive. Discrimination between different types of defects is not good even when checked using different thermal techniques.

surfaces or the presence of scratches. Similarly, delaminations or cracks normal to the ultrasonic wave beam will be detected. Commonly, frequencies of 1–10 MHz are used, for which cracks parallel to the beam or small defects of any nature are unlikely to be seen. Van Dreumel[13] claimed that fibers could be seen and alignment judged, but this probably occurred due to the bunching of fibers or to the diffraction of the waves around bundles of fibers. Small voids cannot be detected unless they appear in large numbers; large voids will be found when their size approaches the wavelength of the probing wave.

Mool and Stephenson[14] made a composite panel comprising 11 layers of boron/epoxy tape sandwiched between aluminum alloy skins. Defects were introduced into the specimen (delaminations, missing plies, extra plies, misaligned filaments, etc.). Commercial through-transmission C-scan equipment was used, which included 5 MHz flat transducers, a long-focus 10 MHz transducer, and a short-focus 15 MHz transducer. The tests were successful—all of the artificial defects being positively located, as well as other defects that were confirmed destructively. The higher-frequency transducers gave sharper defect definition than the lower-frequency transducers, but the attenuation was very much higher. No indication was given of the techniques for identifying an unknown defect.

Liber et al.[12] detailed C-scan equipment and test results from cross-ply graphite/epoxy test specimens that underwent cyclic loading. Difficulties associated with the detection of standard defects were discussed; initial flaws, designed to simulate in-service flaws, were introduced to the test specimens. A "natural standard for gapless delamination (unbonded but contacting areas of adjacent plies)" was obtained from specimens with drilled holes that were known to contain delaminations extending from the boundary of the hole to the interior of the specimen. For surface defects, it was sometimes observed that, in successive scans, the flaw indication reduced in size; this was traced to water penetration and was largely overcome by sealing the edges. Liber et al.[12] adopted a realistic approach to flaw-type discrimination, stressing the need to interpret ultrasonic indications on the basis of previous experience and knowledge of material.

Additional NDI Methods

Whitcomb[15] described the fatigue process in composite material as a complicated combination of "matrix crazing, delamination, fiber failure, fiber/matrix interfacial bond failure, void growth and cracking" and decided that it was "difficult or impossible to handle using traditional NDE techniques." Clearly, better (or at least different) techniques are needed and many alternatives have been proposed, as listed in Table 11.2.

In recent years, neutron radiography has received much attention as a possible new and exciting NDI technique, to complement ultrasonic and x-ray techniques; it is particularly well suited to examining bond lines and to studying composite material in close proximity to metal. Hydrogen, boron, and gadolinium exhibit neutron absorption coefficients "2 or 3 orders of magnitude greater than the average value of structural metals."[16] Thus, organic materials such as epoxy adhesives, which contain 8–12% of hydrogen, exhibit good radiographic contrast; for constant material thickness, variations in film density indicate variations in absorber uniformity caused by voids, inclusions, or material inhomogeneities. There may be difficulties in interpretation, e.g., a low-absorption inclusion looks like a void, but high-absorption inclusions are readily recognized. The neutron radiography technique has been developed using thermal neutrons (i.e., fast neutrons moderated by water or polyethylene) from a nuclear reactor. In-service application awaits the development of a suitable safe, cost-effective portable neutron source.

Although resin is nonconductive, there is a measurable conductivity associated with bundles of graphite fibers and eddy current measurements can be used to determine resin content. (For practical applications, present methods for determining resin content are all based on a time-consuming acid-digestion scheme.) Eddy current tests were made on a graphite/epoxy combination[8] using test frequencies of 0.5–3 MHz and coil diameters of 1–3 mm (depending on specimen thickness). Good correlation between resin content and a chosen eddy current parameter (measured on a phase diagram) was obtained. Owston[17] was confident that eddy currents could be used in various ways (for crack detection, volume fraction, and lay-up order measurement), provided the frequencies could be sufficiently increased. Encouraging results were obtained at 25 MHz and development along these lines will be followed with interest.

Various optical holographic techniques are listed in Table 11.2. One technique involves the measurement of speckle patterns. Speckles appear everywhere in space when a surface is illuminated with laser (coherent) light. The deformation of a structure by mechanical or thermal means modifies the speckle pattern by adding localized regions of high fringe density, which are likely to indicate the presence of a flaw. Another technique is the "live fringe" technique in which a hologram of an unstressed object is recorded and is processed in situ or very accurately replaced. An interferometric comparison is then made by looking through the hologram at the object; fringes are observed when the latter is slightly strained. The live fringe technique of holographic interferometry produces a measure of changes in surface displacement.

Marchant[18] examined Harrier graphite/epoxy wing tips (approximately $2 \times 1 \times 0.1$ m) using live fringes formed from an argon ion lamp. The wing tips were mounted on a heavy steel table that was isolated from ground-induced vibration. Minor problems were experienced with airborne vibration. Exposure times of about 1 s were used and reasonable hologram quality was obtained. Specimens were stressed by heating with a domestic radiator and, although uneven heating occurred, it always seemed possible to make the indications of suspect areas reappear with repeated loading. Good agreement with the radiographic tests was obtained, but doubt was expressed concerning the nature of the defect. It would appear from Marchant's work that, although the technique holds considerable promise, extensive development is still required. Using a thermally induced frozen fringe technique, Daniel and Liber[19] show excellent photographs of fringe patterns, but their discussion is mostly in very general terms. It is not clear to what extent differences between different types of defects can be detected.

Commercial equipment for acoustic holography/imaging frequently includes C-scan as an option. Both amplitude and phase are used to record the hologram. Sheldon[20] used an imaging system in which the scan information was stored in a memory. On command, the various configurations could be called up. Using a 5 MHz focused transducer, 12 and 25 mm debonds between graphite/epoxy skin and honeycomb core were located on the reconstructed C-scan. A focused image technique used a single focused transducer imaging on the back side of the test sheet. A crack in a graphite/epoxy wing attachment trunnion was found by imaging and confirmed by C-scan. Damage specimens were also identified using both techniques. It was generally found that the indicated areas of impact damage were larger than those confirmed visually, i.e., the damage could well have been more extensive than expected.

For many years thermography has been presented as an NDI technique having great potential—very rapid scanning of large surface areas is possible and equipment can be some distance away from the test surface. Consequently, examination can be shifted around to monitor more than one problem area. Failure to realize the potential of thermography undoubtedly stems from the problems attendant on a thermal technique, such as the effects of draughts, variations in surface emissivity, etc., all of which are difficult to overcome, particularly in the field. Other problem areas are identified in Table 11.2.

There are two types of thermal field in materials[21]: (1) stress-generated thermal fields (SGTF) that appear as a consequence of cyclic loading and wherein maximum temperature rises can be expected where stresses are highest, e.g., around flaws; and (2) externally applied thermal fields (EATF) where normally uniform isotherms are distorted in the presence of a flaw or a damaged region. Thermography is the science of measuring temperature change arising from these two thermal fields.

Henneke et al.[22] demonstrated the use of SGTF by cyclic loading of a methyl methacrylate specimen containing a central hole; an isotherm pattern closely related to the calculated stress field was obtained. Furthermore, very good agreement between predicted and measured temperatures was obtained. Cycling loading tests at frequencies between 15 and 45 Hz were conducted on boron/epoxy specimens containing a central hole. Early in testing, heat patterns developed (around the holes) that appeared to be related to stress fields; subsequent changes in the patterns were attributed to the development of fatigue damage. Similar successful tests were conducted on graphite/epoxy laminates containing notches from which matrix cracks propagated. McLaughlin et al.[21] were less successful, probably because their test frequencies were much lower (0.5–5 Hz). Temperature rises were observed for glass/epoxy specimens containing a part-through hole after only 30 cycles at 1 Hz and at only 10% of the static failure load of the flawed specimen. No changes were observed for graphite/epoxy material after 1000 cycles at up to 30% of the static ultimate load. Thus, there appear to be limiting loading frequencies below which no observable change can be expected.

Thermography is an NDI technique that possesses potential, but much development is needed. Sensitivities commonly quoted are about 0.1 °C, temperature ranges are a few degrees Celsius, and the detectable defect size is a few millimeters in diameter. There are many restrictions on the technique and its application to composite material, but delamination is reasonably easy to detect. The success rate seems to vary considerably. Thermography is unlikely to give any information not found with ultrasonic C-scan, but it is a noncontact technique that can be used at a distance and crack growth can be monitored as it occurs.

11.3 ATTEMPTS TO ASSESS STRUCTURAL INTEGRITY

It should be clear that the techniques just described are suitable for detecting and locating a restricted range of defects in composite materials. Defects differ for different types of composite material, but a general classification based on detectability is not hard to arrange. However, because of presently incomplete understanding of failure modes, inspections tell very little about the defect's severity and even less about the life to failure.

Of far greater importance than finding defects is the need to appreciate their importance, i.e., to develop a failure predictor or life indicator. In this section, the measurement techniques described in Table 11.3 will be discussed with this requirement in mind. It will be seen that most go at least part of the way toward this goal.

Vibration Measurements

The use of a vibration technique to locate defects in structures made from advanced composite materials has been proposed.[23] Damage can be detected, located, and roughly quantified by measuring changes in the natural frequencies of the structure. It is claimed that the severity of the damage can be assessed by additional analysis. This technique is potentially very attractive because properties can be measured at a single point on a structure and hence access to the whole of a structure is not required. Actual test time can be very small, particularly if the resonant frequencies are excited by an impulse. However, it is necessary to conduct tests on composites in a constant (± 1 °C) temperature enclosure.

Cawley and Adams[23] readily located damage by saw cuts in a graphite/epoxy plate. One side of a similar plate was damaged by a steel ball; the damage was successfully located by vibration techniques and was confirmed by ultrasonic measurements.

Adams and Flitcroft[24] showed that matrix and interface cracking in graphite or glass/fiber reinforced composite material could be detected in the laboratory using a resonant torsion pendulum. The specific damping capacity was measured from the power needed to maintain a constant vibration amplitude at resonance, while the shear modulus was found from the resonant frequency. Crack size was reliably indicated by the amplitude dependence of these dynamic properties of which damping was the more sensitive measure.

Sims et al.[25] were more interested in evaluating specimen life rather than determining the presence of defects. Most of their results were obtained on

Table 11.3 Attempts to Assess

Technique	Operating Principles
Vibration measurements	A structure or specimen is vibrated through a suitable range of frequencies. Variations in resonant frequencies and amplitudes at or around resonance are measured, from which can be computed the location and severity of damage.
Ultrasonic attenuation	An ultrasonic pulse is injected into a specimen. Attenuation is measured by comparing the initial magnitude of the pulse with that after reflection/transmission in a specimen.
Stress wave factor	An empirical factor proposed by Vary that purports to measure the efficiency of energy transfer by the tested region of material lying between two transducers. The transmitter injects an ultrasonic pulse into the specimen. The pulse propagates through the specimen, is detected, and is processed as the number of crossings above a preset voltage threshold.
Acoustic emission	Elastic waves resulting from deformation or fracture are detected on the surface of the specimen using a sensitive transducer. Electrical signals from the transducer are conditioned and displayed in various ways.

0°/90° cross-ply laminate glass/epoxy material. Complex dynamic moduli and damping factors were determined using a simple resonance technique as well as a torsion pendulum technique. For all systems, the dynamic moduli decreased while the loss factors increased with the introduction of damage. It was concluded that the energy dissipated per cycle by the cracks during dynamic testing was proportional to the total crack area.

Measurement of Ultrasonic Attenuation

In any C-scan measurement, the effects of ultrasonic attenuation are evident. However, attenuation is an ultrasonic parameter of value in its own right, measurement of which can be used to assess structural integrity.

Saluja and Henneke[26] claimed, and were able to justify, that transverse cracks which develop in the weakest plies tend to attain a uniform, equilibrium spacing. These cracks diffract acoustic waves, giving rise to a wave

Structural Integrity

Capabilities	Remarks
Appears to possess potential for determining damage without regard to nature of damage. It is claimed that the severity of damage can also be measured.	Not fully developed. Access to only one point on a structure is frequently sufficient to permit testing to be conducted.
Ultimate strength of lab specimens has been found to correlate with initial attenuation. Measurements of attenuation caused by the formation of a network of cracks in matrix material have identified structural changes.	Attenuation appears to be frequently dependent, but largely independent of life. Access for transducer placement is required (as for other acoustic methods).
Eventual failure sites in laboratory specimens can be identified.	A highly empirical technique that nonetheless appears to be successful in restricted applications.
Differences in time of arrival of signals at an array of transducers permits signal location; there are difficulties with some materials for which acoustic properties vary with direction. Fiber fracture can be distinguished from matrix failure by studying amplitude distributions. Successful safe life predictions have been made in specific cases from proof test measurements.	Retains potential as a failure predictor but has only been successfully used in special cases.

attenuation. Attenuation was claimed to give a good indication of damage; it varied with changes in the crack opening for a fixed number of cracks, was sensitive to frequency, and was likely to depend on the number of cracks for a given constant crack opening. Unlike many other workers, Saluja and Henneke confirmed their findings by destructive examination. Hagemaier and Fassbender[8] found attenuation (which was frequency dependent) correlated well with the void content in simple graphite/epoxy laminates. Unfortunately, no correction was made for specimen thickness (number of plies), which turned out to be another variable. Williams and Doll[27] measured attenuation at intervals of 3×10^4 cycles during a compression-compression fatigue test on graphite/epoxy composite material. There appeared to be a correlation between initial attenuation and cycles to fracture, which improved with increasing test frequency.

Attenuation, simply determined, appears to be sensitive to hygrothermal effects for glass/epoxy composites, but not for graphite/epoxy composites.[28] Accompanying the increased attenuation in the former material is a drastic reduction in flexural strength. However, for both materials (and with the Kevlar material), good correlation between changes in normalized strength and attenuation was found, although no real indication was given for strength reductions greater than 30%. Clearly, confusion can well arise from the attenuation results unless it can be shown that degradation from various processes arises from the same physical phenomena, which seems unlikely.

Stress Wave Factor

The concept of a stress wave factor was developed by Vary and Bowles.[29] They studied the interrelation between the various parameters that influence the strength of a unidirectional graphite/polyimide composite. On the basis of their measurements, Vary and Bowles derived the concept of a stress wave factor. This factor was determined by injecting a repetitive ultrasonic pulse into a specimen using a broadband transducer and detecting the resulting signal some distance away by means of a resonant transducer. The two sensors could obviously be located in various ways. The stress wave factor ε was defined by $\varepsilon = grn$, where g is the period over which measurement is made, r the repetition rate of the input pulse, and n the ring down counts per burst.

Vary and Bowles claimed to be able to predict the relative mechanical strength of a composite material by means of ultrasonic-acoustic measurements made within a relatively "narrow frequency domain" (0.1–2.5 MHz) and without the need for sophisticated equipment. The stress wave factor was shown to correlate strongly with interlaminar shear strength for the particular material. There is no detailed physical basis for any of this work, but it is clear that physical properties should be determinable from a study of wave propagation.

Later work by Vary and Lark[30] deals with more specific NDI applications of the stress factor approach. During the scanning of tensile specimens of graphite/epoxy composite prior to a test, minimum values of the stress factor were observed at a few positions along the specimen. After testing, it was confirmed that failure occurred only at the previously indicated positions. It was claimed that stress wave factor "may be described as a measure of the efficiency of stress wave energy transmission" in a given composite. Hence, it was a sensitive indicator of strength variations and could aid in predicting potential failure locations.

Acoustic Emission

Acoustic emission (AE) is defined by ASTM 610-77[31] as "the class of phenomena whereby transient elastic waves are generated by the rapid release of energy from a localized source or sources within a material, or the transient wave(s) so generated." These waves propagate through a structure and are usually detected by a piezoelectric transducer. The resulting electrical signals can then be processed in various ways to give a wide variety of parameters.[32] AE signal analysis has the potential not only to locate sources and thereafter to define defects in a structure, but also to monitor structural integrity during proof testing and in service. However, the case of a single source in a "simple" material was only recently addressed by Hsu and Eitzen.[33] In practice, the deconvolution of the detected signals into a precise measure of the source function is a complex problem even for relatively uncomplicated metal structures.

There are additional problems that must be solved before AE can be used for routine monitoring of the structural integrity of composite materials; some of these problems have been detailed in the reviews by Williams and Lee[34] and Duke and Henneke.[35]

There are numerous specific mechanisms that produce AE in composite materials,[36] e.g., fiber fracture, matrix cracking, delamination, etc., many of which have already been mentioned. Wave propagation characteristics are complex and are dependent on composite type and design. Signal modification occurs during wave propagation due to the geometric spreading of the wave, the effects of structural boundaries, the frequency dependence of the attenuation, and the anisotropic nature of the composite material. The effects of all these phenomena must be considered during signal analysis. Finally, we need to know the relationship between the AE parameter and the structural integrity of the component.

Several authors have reported success in the use of amplitude distributions to distinguish AE sources and hence identify failure modes.[37,38]

In selected special situations, AE has been used successfully for monitoring structural integrity, but it is far from viable as a universal method. In an early application, Wadin[39] described how AE counts, measured during a proof test, were used to predict impending failure of the fiberglass boom of an aerial lift device. His flaw predictions were confirmed destructively. Fowler[40] and Fowler and Gray[41] developed acceptance-rejection criteria for fiberglass tanks, pressure vessels, and piping based on laboratory and fatigue tests. They introduced the Felicity ratio defined, during repeated loading, as the load at the onset of AE divided by the maximum load previously attained. Their criteria are based on a combination of total counts, signal amplitude, AE activity during a load hold, and the Felicity ratio.

Hamstad[36] and Wadin[42] discussed the presence or absence of the Kaiser effect (defined as the lack of detectable AE until previously reached stress levels are exceeded) in composite materials in terms of the viscoelastic matrix. Deformation at any stress level is significantly time dependent, resulting in time-dependent AE. Thus, the absence of a Kaiser effect allows the determination of a Felicity ratio (as observed by Fowler). This ratio, in conjunction with observed AE, can be used to assess structural integrity. Bailey et al.[43] used a similar approach to assess impact damage.

In addition to the above, many papers deal primarily with "data gathering." Future research will need to concentrate on developing a suitable universal model to describe composite material behavior, before the potential of AE as an indicator of structural integrity can be fully realized.

11.4 SUMMARY

Detection of manufacturing or in-service defects in composites can be accomplished using NDI, especially ultrasonic C-scan and radiography. However, establishing the significance of defects remains a major problem. Four contrasting techniques have been proposed for the assessment of structural integrity:

(1) Vibration measurements, from which damage (of any nature) can be located and a measure of the damage severity can be obtained.

(2) Ultrasonic attenuation, in which changes can be related to the damage rather than the individual defects. From measurements of initial attenuation, failure loads or cycles to failure could be predicted.

(3) Stress wave factor, which is a measure of energy transmission and is essentially a bulk ultrasonic parameter, enabling potential failure sites to be predicted.

(4) Acoustic emission, which is presently only confirmed as a failure predictor derived on the basis of a series of tests at various loads (and, in some cases, loading a component to failure).

None of these candidate techniques is entirely satisfactory. Attenuation and stress wave factor techniques appear to have little scope for future development. Both vibration and acoustic emission techniques appear to possess the potential for predicting failure, although considerable research and development is required. However, until this research is undertaken, it will still be necessary to have recourse to the traditional C-scan and x-radiographic techniques.

References

[1]Scott, I. G. and Scala, C. M., "A Review of Non-destructive Testing of Composite Materials," *Non-Destructive Testing International*, Vol. 15, 1982, pp. 75–86.

[2]Harris, B., "Accumulation of Damage and Non-destructive Testing of Composite Materials and Structures," *Annales de Chimie—Science de Materiaux*, Vol. 5, 1980, pp. 327–339.

[3]Prakash, R., "Non-destructive Testing of Composites," *Composites*, Vol. 11, 1980, pp. 217–224.

[4]Salkind, M. J., "Early Detection of Fatigue Damage in Composite Materials," *Journal of Aircraft*, Vol. 13, 1976, pp. 764–769.

[5]Nevadunsky, J. J., Lucas, J. J., and Salkind, M. J., "Early Fatigue Damage Detection in Composite Materials," *Journal of Composite Materials*, Vol. 9, 1975, pp. 394–408.

[6]Martin, B. G., "Analysis of Radiographic Techniques for Measuring Resin Content in Graphite Fiber Reinforced Epoxy Resin Composites," *Materials Evaluation*, Vol. 35, Sept. 1977, pp. 65–68.

[7]Forli, D. and Torp, S., "NDT of Glass Fiber Reinforced Plastics (GRP)," Paper 4B2, Eighth World Conference on NDT, Cannes, France, 1976.

[8] Hagemaier, D. J. and Fassbender, R. H., "Non-destructive Testing of Advanced Composites," *Materials Evaluation*, Vol. 37, June 1979, pp. 43–49.

[9] Roderick, G. L. and Whitcomb, J. D., "X-ray Method Shows Fibers Fail During Fatigue of Boron/Epoxy Laminates," *Journal of Composite Materials*, Vol. 9, 1975, pp. 391–393.

[10] Crane, R. L., Chang, F. F., and Allinikov, S., "Use of Radiographically Opaque Fibers to Aid the Inspection of Composites," *Materials Evaluation*, Vol. 36, Sept. 1978, pp. 69–71.

[11] Crane, R. L., "Measurement of Composite Ply Orientation Using a Radiographic Fringe Technique," *Materials Evaluation*, Vol. 34, April 1976, pp. 79–80.

[12] Liber, T., Daniel, I. M., and Schraum, S. W., "Ultrasonic Techniques for Inspecting Flat and Cylindrical Composite Cylinders," ASTM STP 696, 1979, pp. 5–25.

[13] Van Dreumel, W. H. M., "Ultrasonic Scanning for Quality Control of Advanced Fiber Composites," *Non-Destructive Testing International*, Vol. 11, 1978, pp. 233–235.

[14] Mool, D. and Stephenson, R., "Ultrasonic Inspection of a Boron/Epoxy-Aluminum Composite Panel," *Materials Evaluation*, Vol. 29, 1971, pp. 159–164.

[15] Whitcomb, J. D., "Thermographic Measurement of Fatigue Damage," ASTM STP 674, 1979, pp. 502–516.

[16] Dance, W. E. and Middlebrook, J. B., "Neutron Radiographic Non-destructive Inspection in Bonded Composite Structures," ASTM STP 696, 1979, pp. 57–71.

[17] Owston, C. N., "Eddy Current Methods for the Examination of Carbon Fibre Reinforced Epoxy Resins," *Materials Evaluation*, Vol. 34, 1976, pp. 237–244, 250.

[18] Marchant, M., "Holographic Interferometry of CFRP Wing Tips," Royal Aeronautic Establishment, TR-78105, Aug. 1978.

[19] Daniel, I. M. and Liber, T., "Non-destructive Evaluation Techniques for Composite Materials," *Proceedings of 12th Symposium on NDE, ASNT and NTIAC*, San Antonio, TX, April 1979, pp. 226–244.

[20] Sheldon, W. H., "Comparative Evaluation of Potential NDE Techniques for Inspection of Advanced Composite Structures," *Materials Evaluation*, Vol. 36, 1978, pp. 41–46.

[21] McLaughlin, P. V., McAssey, E. V., and Deitrich, R. C., "Non-destructive Examination of Fibre Composite Structures by Thermal Field Techniques," *Non-Destructive Testing International*, Vol. 13, 1980, pp. 56–62.

[22]Henneke, E. G., Reifsnider, K. L., and Stinchcomb, W. W., "Thermography—An NDI Method for Damage Detection," *Journal of Metals*, Vol. 31, Sept. 1979, pp. 11–15.

[23]Cawley, P. and Adams, R. D., "Vibration Technique for Non-destructive Testing of Fibre Composite Structures," *Journal of Composite Materials*, Vol. 13, 1979, pp. 161–175.

[24]Adams, R. D. and Flitcroft, J. E., "Assessment of Matrix and Interface Damage in High Performance Fibre Reinforced Composites," Paper 4B3, Eighth World Conference on NDT, 1976.

[25]Sims, G. D., Dean, G. D., Read, B. E., and Western, B. C., Assessment of Damage in GRP Laminates by Stress Wave Emission and Dynamic Mechanical Measurements," *Journal of Materials Science*, Vol. 12, 1977, pp. 2329–2342.

[26]Saluja, H. S. and Henneke, E. G., "Ultrasonic Attenuation Measurement of Fatigue Damage in Graphite/Epoxy Composite Laminates," *Proceedings of 12th Symposium on NDE, ASNT and NTIAC*, San Antonio, TX, April 1979, pp. 260–268.

[27]Williams, J. H. and Doll, B., "Ultrasonic Attenuation as an Indicator of Fatigue Life of Graphite/Epoxy Fiber Composite," NASA CR 3179, 1979.

[28]Bar-Cohen, Y., Meron, M., and Ishai, O., "Non-destructive Evaluation of Hygrothermal Effects on Fiber-Reinforced Plastic Laminates," *Journal of Testing and Evaluation*, Vol. 7, 1979, pp. 291–296.

[29]Vary, A. and Bowles, K. J., "Ultrasonic Evaluation of the Strength of Unidirectional Graphite-Polyimide Composite," NASA TM X-73646, 1979.

[30]Vary, A. and Lark, R. F., "Correlation of Fiber Composite Tensile Strength with the Ultrasonic Stress Wave Factor," *Journal of Testing and Evaluation*, Vol. 7, 1979, pp. 185–191.

[31]*Annual Book of ASTM Standards, Part 11: Metallography: Nondestructive Testing*, ASTM, Philadelphia, 1977, p. 676.

[32]Licht, T., "Acoustic Emission," *Bruel & Kjaer Technical Review*, Vol. 2, 1979.

[33]Hsu, N. N. and Eitzen, D. G., "AE Signal Analysis—Laboratory Experiments Examining the Physical Processes of Acoustic Emission," *Proceedings of Fifth International Acoustic Emission Symposium*, Tokyo, Nov. 1980, pp. 67–78.

[34]Williams, J. H. and Lee, S. S., "Acoustic Emission Monitoring of Fiber Composite Materials and Structures," *Journal of Composite Materials*, Vol. 12, 1978, pp. 348–370.

[35]Duke, J. C. and Henneke, E. G., "Acoustic Emission Monitoring of Advanced Fiber Reinforced Composite Materials," *Proceedings of Fifth International Acoustic Emission Symposium*, Tokyo, Nov. 1980, pp. 147–162.

[36]Hamstad, M. A., "Deformation and Failure Information from Composite Materials via Acoustic Emission," *Fundamentals of Acoustic Emission*, edited by K. Ono, School of Engineering and Applied Science, University of California, Los Angeles, CA, 1979, pp. 229–260.

[37]Bailey, C. D., Freeman, S. M., and Hamilton, J. M., "Acoustic Emission Monitors Damage Progression in Graphite Epoxy Composite Structures," *Materials Evaluation*, Vol. 38, Aug. 1980, pp. 21–27.

[38]Rotem, A., "The Discrimination of Micro-Fracture Mode of Fibrous Composite Material by Acoustic Emission Technique," *Fibre Science and Technology*, Vol. 10, 1977, pp. 101–121.

[39]Wadin, J. R., "Listening to Cherry Picker Booms," Dunegan/Endevco, San Juan Capistrano, CA, Rept. LT2, March 1977.

[40]Fowler, T. J., Acoustic Emission of Fiber Reinforced Plastics, ASCE Paper 3092, Oct. 1977.

[41]Fowler, T. J. and Gray, E., "Development of an Acoustic Emission Test for FRP Equipment," ASCE Paper 3583, April 1979.

[42]Wadin, J. R., "Listening to Composite Materials," Dunegan/Endevco, San Juan Capistrano, CA, Rept. LT4, Aug. 1978.

[43]Bailey, C. D., Hamilton, J. M., and Pless, W. M., "Acoustic Emission of Impact-Damaged Graphite/Epoxy Composites," *Materials Evaluation*, Vol. 37, May 1979, pp. 43–48, 54.

12. REPAIR OF GRAPHITE/EPOXY COMPOSITES

12.1 INTRODUCTION

Graphite/epoxy composites have many advantages for use as aircraft structural materials, including their high specific strength and stiffness, resistance to damage by fatigue loading, and immunity to corrosion. Thus, extensive use of these composites should reduce the high maintenance costs associated with repair of corrosion damage normally encountered with conventional aluminum alloys, particularly those exposed in a marine environment. Similarly, costs associated with repair of damage due to fatigue should also be substantially reduced, since the composites do not, in general, suffer from the cracking encountered with metallic structures, particularly cracking initiating from fretting damage in fastener holes or from corrosion pitting.

However, maintenance costs associated with repair of service contact damage is expected to increase, since graphite/epoxy is essentially an unforgiving brittle material—unable to yield plastically under overload. Even quite modest impacts (by metallic standards) can lead to internal damage in the form of delaminations, which may result in a marked strength reduction, particularly under compression loading. The impacted area may not be apparent from surface examination because of the absence of permanent deformation.

Other more severe handling and environmental damage will also occur; however, this is also common to metallic structures, particularly those of honeycomb construction.

It is the purpose of this chapter to consider the repair of graphite/epoxy aircraft components of various configurations. Typical aircraft component forms are listed in Table 12.1. Repairs described here vary from the simple injection of resins into small delaminations and disbonds to a variety of patching procedures for damage to regions extending up to 100 mm or so. Emphasis will be on repairs that can be carried out under field or depot conditions; however, some of the repairs described will be more appropriate to factory conditions. Many of the repairs for composite faced sandwich panels are similar to those described for metallic honeycomb structures in Refs. 1 and 2. The following section gives a brief discussion of some of the salient factors concerning damage, inspection, and repair criteria. The rest of the chapter is concerned with repair methodology.

12.2 DAMAGE ASSESSMENT

Types of Defects

Defects may be present initially in the structure due to faulty manufacture or may be introduced during service due to damage resulting from mechanical contact or environmental effects. Table 12.2 lists manufacturing faults, mechanical contact damage, and service environment damage and provides some details on the source of each type of damage. Although it is not the intention here to concentrate on the repair of manufacturing damage, some mention of this will be made for completeness.

Inspection Procedures

Field procedures. Visual examination can identify most of the severe forms of damage. However, as mentioned previously, external damage associated with internal delaminations may not be visible (other than

Table 12.1 Typical Graphite/Epoxy Aircraft Structure; Ply Configuration Generally of the ±45°/0°/90° Variety[a]

Structure	Typical No. of Plies	Applications
Honeycomb panels: graphite/epoxy skins; aluminum, fiberglass, or nomex core	2–16	Control surfaces, fairings, access doors, flooring, horizontal tail, speed brake
Sandwich panels: graphite/epoxy skins; PVC foam core	2–6	As above
Stiffened panels: graphite/epoxy skins with integral graphite/epoxy stiffeners	16–20	Fuselage shells, tail skins, wing panels
Monolithic panels: graphite/epoxy skins bolted to aluminum alloy or titanium substructure	100^{+}	Main torque box, wing, and tail
Monolithic panels: graphite/epoxy skins bolted to graphite/epoxy substructure	100^{+}	Main torque box
Channels, beams: graphite/epoxy	16–60	Spars (including sine wave spars), ribs

[a] Ply thickness is usually about 0.13 mm.

Table 12.2 Types of Typical Structural Defects and / or Damage

Defect	Typical Causes
	Manufacturing defects
Voids	Poor process control
Delaminations	Inclusion of release film
	Poor process control
	Faulty hole formation procedures
Disbonds (in bonded joints and honeycomb panels)	Poor fit of parts
	Inclusion of release film
	Poor process control
Surface damage	Poor release procedure
	Bad handling
Misdrilled holes	Faulty jigging
	Service mechanical damage
Cuts, scratches	Mishandling
Abrasion	Rain/grit erosion
Delaminations	Impact damage
Disbonds	Impact damage
	Overload
Hole elongation	Overload/bearing failure
Dents (with delaminations and crushed core)	Impact damage
	Walk-in no-step regions
	Runway stones
Edge damage	Mishandling of doors and removable parts
Penetration	Battle damage
	Severe mishandling (e.g., fork lift)
	Environmental damage
Surface oxidation	Lightning strike
	Overheat
	Battle damage (e.g., laser)
Delamination	Freeze/thaw stressing (due to moisture expansion)
	Thermal spike (causing steam formation)
Disbonds (in honeycomb panels)	As for delaminations
	Inadequate surface treatment
Core corrosion	Moisture penetration into honeycomb
Surface swelling	Use of undesirable solvents (e.g., paint stripper)

possibly as a depression of 0.1 mm or so). The simple coin tap test is very good for detecting delaminations and disbonds in some structures such as honeycomb panels, where the support conditions of the damaged material favor its use. Good areas produce a ringing sound, whereas, due to increased damping, disbonded or delaminated regions emit a low pitch or "dead" sound. The usefulness of the technique is limited by the thickness and damping characteristics of the skin.

Pulse-echo ultrasonic procedures, where it is feasible to employ them, are very effective in detecting delaminations and seriously voided regions. Templates may be employed to enable systematic coverage of a component; automatic ultrasonic scanning procedures for field inspection or large regions are presently under development.

Depot inspection. In addition to the above procedures, a wide range of more sophisticated procedures are available for depot level inspection. The most versatile is the water bath or water-coupled C-scan procedure that automatically maps, in the *xy* plane (i.e., the laminate plane), pulse echo or pulse transmission attenuation. This technique can accommodate very large components and is most effective in detecting voids, delaminations, and disbonds since these lie in the *xy* plane. The other major technique is radiography, which is effective in detecting defects in honeycomb parts such as water entrapment, core corrosion, and core splice separation and defects in plane laminates that lie parallel to the x-ray beam such as foreign body inclusion and cracking. In some applications, an x-ray absorbent fluid (such as di-iodobutane) may be forced into disbonded regions to enhance the x-ray image.

Repair Criteria

A rational basis is required for assessment of the seriousness of flaws in composite components. Three basic decisions are possible: (1) the defect is negligible, in which case it may be disregarded apart, perhaps, for some cosmetic treatment; (2) the defect is not negligible, but the component is repairable; or (3) the component is not (economically) repairable and therefore must be replaced.

The decision-making process must allow for the nature of the stressing and the environment, the cost of repair compared with replacement (including the loss of availability of the aircraft), the reliability and efficiency of the repair, and the consequence of failure. A method comparable to linear elastic fracture mechanics, as employed to assess the seriousness of crack-like flaws in metals, has yet to be fully developed for composites or bonded components. However, in general, crack-like flaws normal to the loading direction do not occur in these materials and the processes of failure are much more diffuse. Delaminations in the case of composites, and disbonds in the case of bonded joints, are the major forms of damage; these can be treated as single or multiple internal cracks, aligned parallel to the surface of the component.

The problem is to assess whether these flaws have reduced the residual strength of the structure below an acceptable level or whether the flaws may grow in service and thus, at some stage, reduce the residual strength below an acceptable level. Under tension loading, the flaws are usually not a serious problem (ignoring possible environmental damage, such as freeze/thaw delamination), since load redistribution can occur and growth is fairly slow; in general, the strain energy release rate does not increase with flaw size. Adhesive-bonded joints, in particular, when designed with sufficient safety margin and, especially, sufficient overlap length, allow extensive load redistribution and are thus highly tolerant of flaws.

However, if peel stresses can develop, which is particularly the situation in compression loading, the flaws can grow very rapidly and result in catastrophic buckling failure. Analytical methods are currently being developed to enable assessment of safe flaw size in composite laminates subject to compression; this topic is discussed in Chap. 10. At the present time, the rule of thumb is that delaminations below about 20–30 mm in diameter will not reduce residual strength, or grow under compression-dominated fatigue loading, when subjected to strains below about 4000 microstrain (which is close to the present ultimate allowable strain) and may therefore be left unrepaired. Larger delaminations, particularly those in critical areas exposed to high compression strains, should be repaired.

12.3 GENERAL REPAIR APPROACHES

Requirements of a Repair Procedure

The ideal requirement of a repair is to restore structural capability permanently, with a minimum reduction in functional capability and a minimum increase in weight, particularly on control surfaces where balance is important. However, implementation of the repair should not require excessive downtime of the aircraft, should not excessively increase the size of the damaged area, and should not require elaborate procedures or tooling. Thus, in practice, some compromise between ideal and practical procedures is required.

Structural restoration (described more fully later) requires that the stiffness and strength are restored to the design allowable values in the operating environment. The functional aspects that must be considered are: installation constraints, aerodynamic acceptability, and surface protection.

Installation constraints that must be considered include: allowance for clearance in mating surfaces, available fastener length, requirements for sealing grooves, etc., and system installation requirements (for instance, for support brackets and mounting provisions).

Aerodynamic acceptability refers to patch repairs applied to the external surface of the aircraft. Where possible, flush repairs should be employed to obtain the optimum aerodynamic surface. However, if external (scab) repair patches are employed, some tapering of the patch is required to ensure minimum disturbance of the airflow. Since a taper rate of about 50 : 1 is usually used at the outer ends of bonded patches for structural applications, this requirement is automatically satisfied. In addition to the patch taper, the fillet of adhesive around the edge of the patch helps to ensure good airflow.

Table 12.3 Major U.S. Repair Programs for Advanced Fiber Composites

Ref.	Company/ Sponsor	Objectives
3	Grumman/ U.S. Air Force	Repair holes up to 75 mm diameter in boron/epoxy laminates (as used in F-14) about 3 mm thick using titanium foil/glass/epoxy bonded patches
4	General Dynamics/ U.S. Air Force	Repairs to various types of damage in graphite/epoxy laminates, joints, and components with a wide range of configurations and thicknesses, using resin injection and bonded graphite/ epoxy or titanium alloy patches
5, 6	Northrop/ U.S. Air Force	Repairs to holes up to 100 mm diameter in graphite/epoxy laminates, thickness up to 6 mm, using bonded graphite/epoxy patches
7	McDonnell-Douglas	Repairs to holes up to 100 mm diameter in graphite/epoxy monolithic skins, thicknesses to 13 mm, using bolted titanium alloy patches

Various protection systems are employed on the surface of the aircraft and these must be restored after installation of the repair; included are the following: (1) external environmental protection, e.g., epoxy or polyurethane paints; (2) lightning strike protection, e.g., electrical conductor mesh (which must be spliced into existing network with conductive material); (3) fuel tank sealant and surface coatings; (4) heat shielding, heat and flame resistant material; (5) surface insulation, such as glass/epoxy to avoid galvanic corrosion of aluminum alloys in contact; and (6) wear-resistant coatings such as aramid, thermoplastic sheets, rubber, steel, or titanium alloy that are employed at leading edges.

Major Repair Research and Development Programs

A number of programs on the repair of graphite/epoxy composites have been sponsored by the U.S. Department of Defense, mostly with the aircraft companies. Some of these are listed in Table 12.3.

Repair Procedures

The repair approaches can be broadly divided into nonpatch, usually for minor defects, and patch, usually for more major defects and damage. However, these procedures may be employed in combination for some types of repair. The two approaches are summarized in Table 12.4.

Table 12.4 Repair Procedures

Procedure	Application
Nonpatch repair procedures for minor damage	
Resin injection	Connected voids Small delaminations Small disbonds
Potting or filling	Minor depressions Skin damage in honeycomb panels Core replacement in honeycomb panels Fastener hole elongation
Heat treatment	Remove entrapped moisture in honeycomb panels Dry out absorbed moisture
Surface coating	Seal honeycomb panels Restore surface protection
Patch repairs for major damage[a]	
Bonded external patch graphite/epoxy (cocured; precured layers; precured) titanium alloy foil	Repairs to skins, particularly on honeycomb panels, up to 16 plies thick. Well suited for field application
Bonded flush patch graphite/epoxy (usually cocured)	Repairs to skins 16 to 100 plies thick, holes up to 100 mm. May be difficult to employ under field conditions
Bolted external patch titanium alloy (usual) aluminum alloy	Repairs to monolithic skins 50–100 plies, holes up to 100 mm. Suited for field applications

[a]All are capable of restoring ultimate strain allowables to the limits of laminate thickness noted.

Repairs to minor structural damage.

(1) Injection repairs. Resin injection repairs are used for minor disbonds and delaminations. The effectiveness of this approach depends on whether the defect arose during manufacture or was due to mechanical damage during service. Due to the local lack of bonding pressure or contamination of the bonding surface, manufacturing flaws have a surface glaze that must be removed to ensure high bond strength. This cannot be achieved for internal surfaces; thus, these repairs are unsatisfactory. In contrast, internal flaws caused by mechanical damage have a surface that can be bonded reasonably effectively, provided contamination has not occurred, for instance, with moisture (which can be removed by drying) or service fluids such as fuel or hydraulic oil.

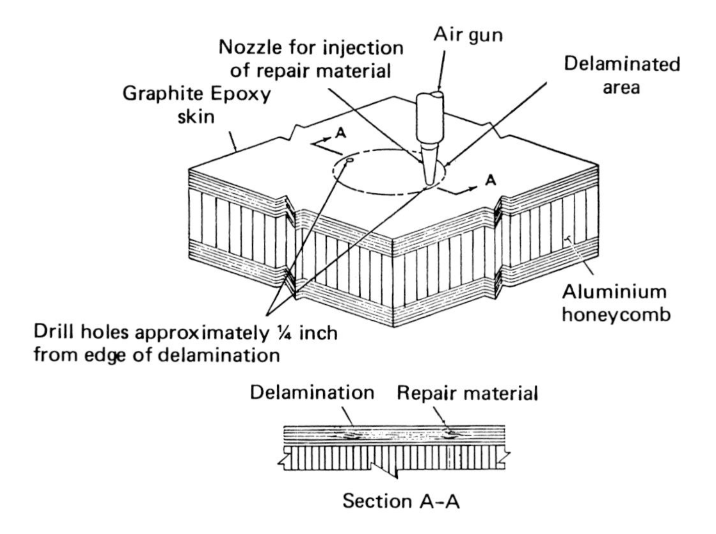

Fig. 12.1 Method of repairing delamination in the composite skin of a honeycomb panel by the resin injection procedure.

The injection procedure, illustrated in Fig. 12.1, involves the formation of several injection and bleeder holes that must penetrate to the depth of the defect; this requires accurate nondestructive inspection (NDI). If the defect is not penetrated, the resin cannot enter the void; if the hole is too deep, damage will occur to the material below the void and the injection may be inefficient. The resin is usually injected after preheating the repair area to about 65°C; prolonged heating may be required to remove moisture. Resin is injected, by means of an air gun, until excess resin flows from adjacent holes. The procedure is repeated until all of the holes have been treated; then the holes are temporarily sealed with a layer of protective tape. Finally, pressure is applied to the repair area to improve the mating of adjacent regions and to improve or maintain contour. The resin is allowed to gel at room temperature, usually followed by a postcure at about 150°C.

Similar procedures can be applied to disbonded joints, provided no corrosion or contamination is present in the damaged region. However, in general, bonded joints have a large strength margin and injection is unlikely to improve strength, since strong bonds will not be obtained. If the disbonds extend to the edge of the bonded region, the procedure is effective in sealing the joint against further damage, e.g., due to the ingress of moisture and consequent corrosion or freeze/thaw delamination. Since the bond strength is not improved, a sealant or soft adhesive may be the best choice for the injection system. If the disbonds do not extend to the edge of the joint and are reasonably small, they are best left alone, because a faulty injection repair would simply provide a path for moisture to the sensitive bond interface.

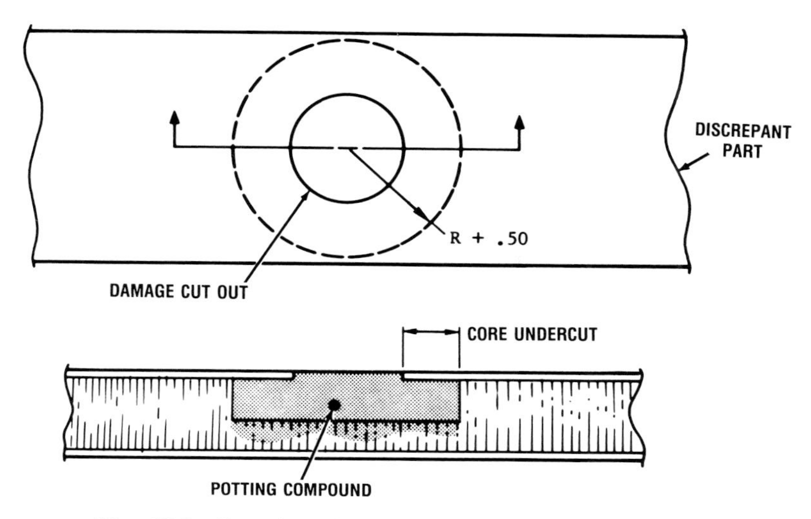

Fig. 12.2 Potted repair to a damaged honeycomb region.

(2) Filler or potting-type repairs. Potting repairs are made by filling the defective region with a filler compound; minor indentations may be filled in this way provided NDI has ensured that no serious internal matrix cracking or delaminations are present. In the case of lightly loaded honeycomb panels, potting repairs may be made to stabilize the skin and seal the damaged region. The repair in this case, as illustrated in Fig. 12.2, involves removal of the damaged skin and core and then further undercutting of the core to ensure mechanical entrapment of the potting compound. The resin in the potting compound is then cured at about 150°C. An alternative is to plug the cavity with glass cloth/epoxy prepreg; however, this imposes a higher weight penalty.

Damage in attachment holes, such as minor hole elongations or wear damage, may be repaired with machinable potting compound. Mislocated or oversize holes can be rectified by filling the hole with either an aluminum rod (adhesively bonded in position) or machinable potting compound and then redrilling.

Patch repairs. Patch repairs are generally employed to repair major damage and essentially involve replacing the lost load path with new material joined to the parent structure. Thus, the repair is best considered as a joint for the purpose of design; Fig. 12.3 illustrates the various bonded joint configurations applicable to patch repairs and the resulting shear stress distribution in the adhesive. This subsection outlines the various repair options; design considerations will be given later.

(1) Bonded external patch repairs. In this approach, the damaged region is removed, leaving a straight (or preferably tapered) hole and over this region a patch with tapered (or stepped) ends is bonded to the parent

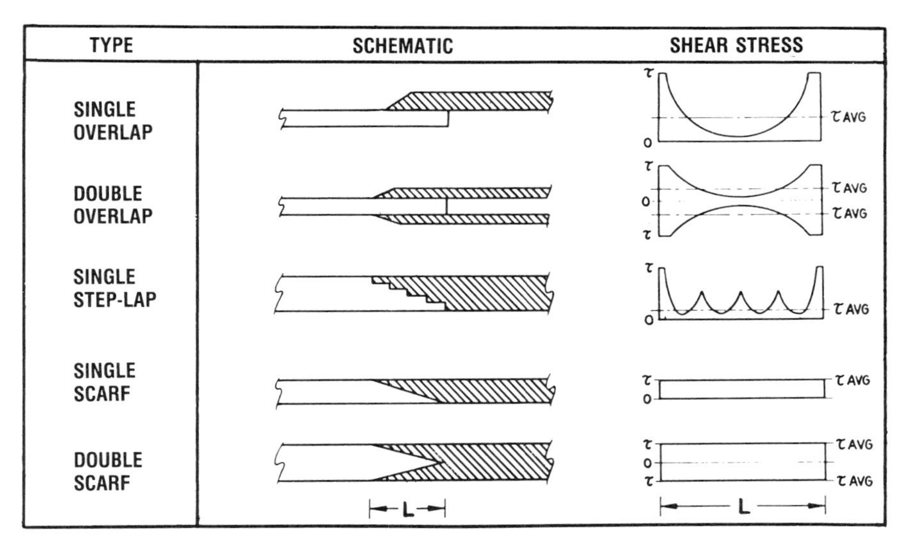

Fig. 12.3 Main types of joint configurations employed for bonded patch repairs and the resulting shear stress distributions.

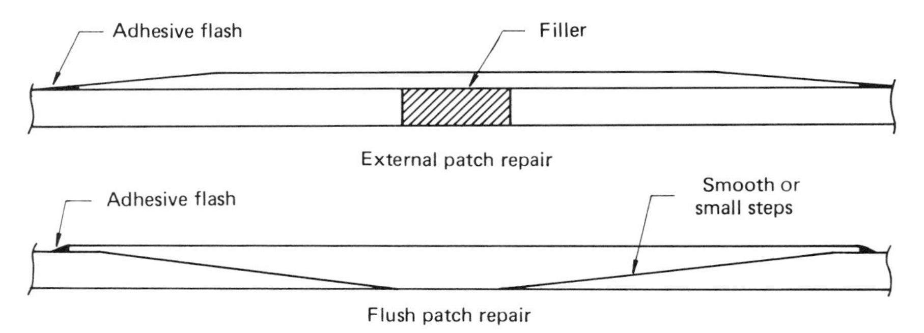

Fig. 12.4 Two main types of patch repair.

laminate (Fig. 12.4). This repair configuration is similar to that of a tapered single-overlap joint (Fig. 12.3); the taper is most important to reduce peel and shear stresses that would otherwise cause failure of the patch. However, a taper is not required for patches of only a few ply thicknesses. External patches can be employed reasonably successfully (depending on the stressing requirements of the area) to repair honeycomb skins of thickness up to about 16 plies. This type of repair will be the most widely employed, since external patches are relatively easily applied under field conditions. Strength recoveries of 50–100% of ultimate allowables of the parent material can be achieved, depending on the laminate thickness.

The main problem with external patches is that, as in a single-lap joint, there is an eccentric load path that results in quite severe bending in the patch and peeling stresses in the adhesive and composite. Out-of-plane bending under compressive axial loads can also significantly reduce the buckling stability. However, these effects are greatly reduced if the patched region is supported by a substructure, such as honeycomb core, that reacts out the bending.

When repairs are made to a honeycomb panel, the damaged core is also removed and the region filled either with potting compound, layers of glass cloth/epoxy prepreg, or new core bonded with a core splice adhesive; use of a honeycomb core is essential in control surfaces to minimize any weight increase. Honeycomb core is generally the preferred insert in all repairs since its use avoids local hard spots, which can result in stress concentrations occurring in the component. The patch is usually bonded to the repair area with a film adhesive, using a vacuum bag pressurization procedure (Fig. 12.5); a heater blanket may be incorporated into the bag assembly to provide the cure temperature.

Several options exist for the patch; it may be made of graphite/epoxy with a similar ply configuration to the parent laminate; graphite/epoxy with a quasi-isotropic lay-up (to reduce the danger of lay-up and application errors), in which case it would require to be thicker than the parent laminate; or titanium foil layers (usually about 0.2 mm thick) adhesively bonded together.

The graphite/epoxy patch may be: (1) formed over the parent laminate from prepreg tape cut to shape and then cocured with the adhesive, (2) precured in layers and bonded to the parent laminate during the repair with interleaved layers of adhesive, or (3) preformed to shape and then bonded to the parent material in a subsequent operation. This last option produces the best patch properties; however, since the preformed patch is not compliant, serious "fit-up" problems may arise on curved surfaces.

The titanium patch is produced by interleaving the foil with the adhesive; usually a layer of glass cloth is also employed between each layer of titanium foil to reduce the shear modulus of the patch and thus minimize

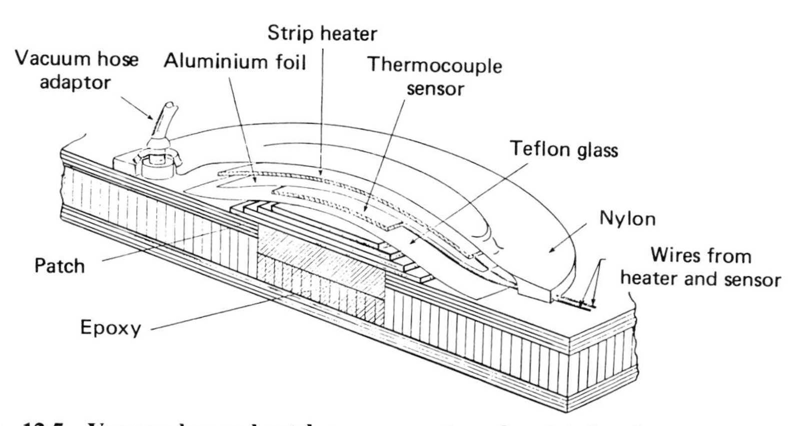

Fig. 12.5 Vacuum bag and patch arrangement employed to bond an external repair patch.

the stress concentrations in the adhesive layer. The main advantages of titanium foil patches are: there is no requirement to control the orientation during formation of the patch and the properties of the patch are not affected by the cure conditions and dissolved moisture in the parent laminate. However, since the titanium patch has a higher thermal coefficient of expansion than the parent material, residual stresses are developed in the adhesive, leading to reduction in the strength of the patch system. A major disadvantage of titanium is that special surface treatment is required prior to adhesive bonding.

(2) Flush patch repairs. The flush patch repair configuration (Fig. 12.4) is similar to a single scarf joint (or a very short overlap single step joint, see Fig. 12.3) and, therefore, has the benefit of a (nearly) uniform shear stress distribution in the adhesive layer. In addition, due to the lack of eccentricity in the load, the patch peel stresses are low. Therefore, flush repairs are highly efficient and are particularly suited to external repairs of thick laminates because of the unlimited thickness of material that can be joined and the smooth surface contour that can be produced.

However, flush repairs are much more difficult and time consuming to apply than external patch repairs and so will usually be employed only under depot or factory conditions. A further, and significant, disadvantage of flush repairs is that they require the removal of a large amount of undamaged material to form the required taper angle—about 18 : 1. For example, taking a fairly extreme (but not unlikely) case of a single-sided flush repair to a 13 mm thick laminate with a 100 mm hole, the scarfed diameter would be nearly 600 mm. However, if a double-sided flush patch could be employed, the tapered zone length would be reduced by about one-half.

Single-sided flush patches can be employed to repair part-through or full-penetration damage. A part-through flush patch, schematically shown in Fig. 12.6, may, as an example, be used to repair a delaminated region in a thick laminate when injection repair is not considered adequate; the material above the delamination is first ground away to leave a recess with the appropriate taper.

Flush repairs are usually based on graphite/epoxy patches with a ply configuration similar to the parent laminate. The graphite/epoxy patches are generally cocured to avoid the severe fit-up problems encountered with precured patches. To cure the patch and adhesive, pressure may be applied by a vacuum bag, heater blanket procedure (Fig. 12.6). Alternatively, in a depot or factory, temperature may be applied in an oven in combination with a vacuum bag pressurization system or, best of all where possible, temperature and pressure may be applied in an autoclave.

In an extensive repair study described in Ref. 5, it was found that peel failure occurred in the longest (outer) 0° fibers in a graphite/epoxy patch, resulting in failure of the patch, unless the ends of the ply were serrated. With a through-thickness repair, it was found to be necessary to scarf the parent laminate to the inner ply. Using this approach, it was possible to produce very efficient repairs to 16 ply laminates with holes.

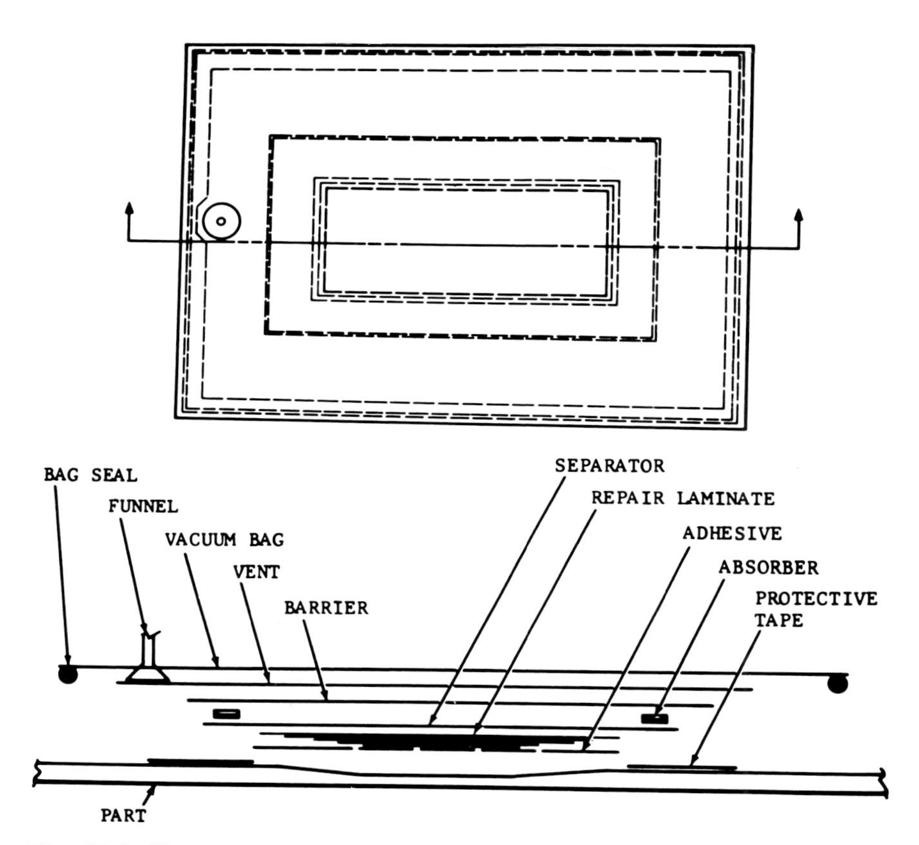

Fig. 12.6 Vacuum bag and patch arrangement employed to bond a flush repair patch.

One-sided flush repairs may be employed to repair disbonded regions in composite-to-metal scarf and stepped-lap joints. Here, it is particularly important to ensure that the metal surface is free from corrosion and correctly treated prior to application of the repair.

(3) External bolted patch repairs. The external bolted patch repair is similar to a bolted single- or double-lap joint, depending on the degree of constraint to bending. A disadvantage of a bolted repair is that quite severe stress concentrations are introduced into a structure at the bolt holes. However, this may not be a significant disadvantage in a mechanically fastened structure. A major advantage of bolted repairs is that the bolts provide a transverse reinforcement with clamping pressure, which is effective in preventing the spread of pre-existing delaminations.

In general, bolted patches are employed for thick laminates (8–15 mm) where the shear stress requirement exceeds the capability of adhesives for external patch repairs and where the complexity and material removal requirements may preclude the use of flush repairs. Moisture problems (discussed below) also limit field applications of bonded repairs in thick materials. Thus, bolted repairs would be employed for field repairs of critical monolithic components such as the F-18 and AV8B Harrier wing skins. Titanium alloy is best employed for the metal patch, nut plate, and fasteners, since (unlike aluminum alloys) it does not suffer galvanic corrosion when in electrical contact with graphite/epoxy and also has a fairly low expansion coefficient, thus minimizing thermal and residual stresses. The type of patch configuration investigated in Ref. 7 is shown schematically in Fig. 12.7; it has proved to be highly effective. The patch has a chamfered edge to minimize any disturbance of the airflow. The nut plate consists of two sections to allow its blind insertion and attachment. Tests described in Ref. 7 have shown that this repair is capable of restoring the strength in 100 ply thick (13 mm) laminates with 100 mm diameter holes to over 4000 microstrain, which is above the usual ultimate design allowable.

(4) External bonded patch repairs to internal structure. Repairs to the substructure (and other complex composite items) may be accomplished using external patches, together with resin injection to diminish the effect of delaminations and disbonds. Although, in some cases, preformed graphite/epoxy patches may be employed, normally the repair would be effected with prepreg materials and cocured in position with adhesive film, using vacuum bagging procedures, as illustrated in Fig. 12.8. The heat source may be radiant heaters, as illustrated, or heat blankets encased in the vacuum bag assembly. Fasteners, may also be employed to reduce peel stresses or reinforce delaminated regions—possibly after resin injection.

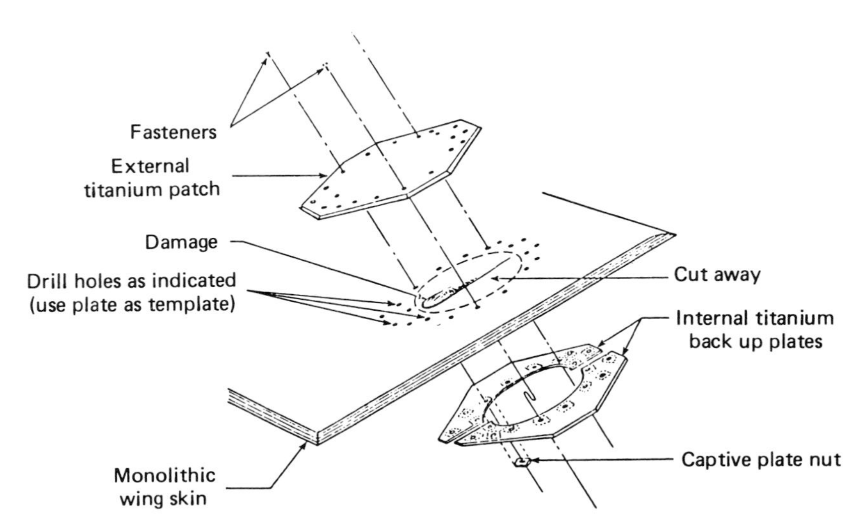

Fig. 12.7 Bolted titanium external patch repair for thick monolithic graphite/epoxy laminates such as employed in aircraft wings.

12.4 MATERIALS ASPECTS

The Moisture Problem

Graphite/epoxy laminates may absorb up to about 1.5% moisture during exposure in humid environments. The moisture is actually absorbed by the epoxy matrix where (by acting as a plasticizer) it reduces some resin-sensitive mechanical properties, such as compressive strength at elevated temperature. Moisture can also be trapped in voids and delaminations where it can cause severe damage due to expansion effects in a thermal spike exposure (due to steam formation) or in a freeze/thaw cycle (due to ice formation) and, in the case of honeycomb panels, where it can cause corrosion of the aluminum alloy core.

Moisture can also cause serious problems during repair implementation if it is not removed by an initial heat treatment. During patch application, the moisture may vaporize and split the laminate in the voided regions, form voids in the adhesive, or form voids in the matrix of the repair laminate (if it is being cocured). Damage in the matrix can be severe if the heat treatment is performed above its glass transition temperature, when its strength is

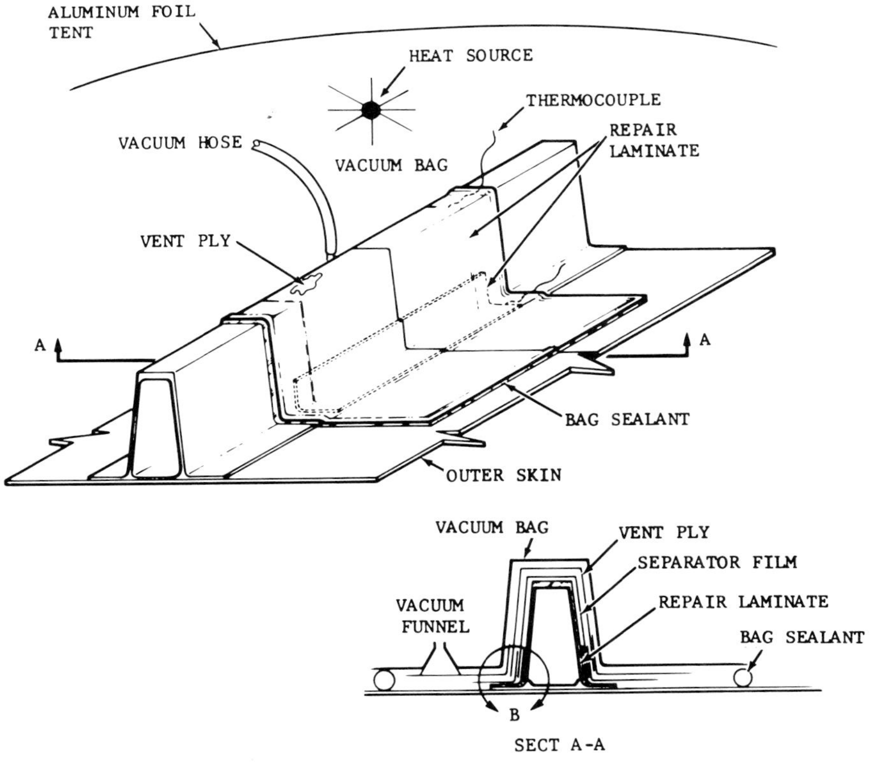

Fig. 12.8 External patch repair to an internal aircraft structure, showing the use of an external heat source.

quite low.[8] During cure of the adhesive or patch, moisture that has diffused in from the parent laminate will produce voids if the partial pressure of the moisture exceeds the applied (hydraulic) pressure during the cure operation.[9] In all cases, the result may be a severe degradation in mechanical properties. The problem of moisture removal is a much more serious problem in a thick laminate (50 plies or more), since days of heating may be required to effect its removal. In general, however, it may not be necessary to remove all of the moisture, since only the surface moisture causes problems in curing the patch and adhesive.

Thin laminates (16 plies or less) can be dried out fairly rapidly. However, if the laminate forms the face of a honeycomb panel, excessive internal steam formation may result in a blown skin. One approach is to accept the presence of moisture in the laminate and allow for it in terms of reduced design allowables—for both the patch (if cocured) and the adhesive. With precured patches, precured plies, or titanium foil patches, the patches do not suffer from moisture problems; however, the problem of adhesive porosity remains the same.

Mechanical Preparation

If a patch repair is to be applied, the damaged region is first outlined in the form of a geometrical shape that allows the accurate preparation and installation of a patch. In general, the shape will be circular and will encompass the area of damage, as determined by NDI and visual inspection, but will include as little as possible of the sound material.

Graphite/epoxy is best cut with tungsten-carbide tipped tools; conventional high-speed tools can be used, but their useful life is unacceptably short. Most forming operations for repair purposes can be performed with an end mill cutter or router mounted in an air motor on a portable base. A template may be used to control the outline of the shape of the cut and shims used to control its depth. Taper cuts may be made, using shims to allow cuts of one ply depth at a time. Alternatively, a sanding drum (alumina or silicon carbide grit) may be used to cut a smooth taper. In this case, the tool may be hand guided (controlled by the operator's observation of the ply exposure) or, preferably, template guided. The taper may extend through the thickness of the laminate in the repair of penetration damage or only part way through in the repair of delamination damage, as shown in Fig. 12.6.

Instead of employing a cutting or grinding operation, it is possible to form the taper using a peeling procedure. This involves employing a sharp knife to cut through the composite, one ply at a time, and then, using grips to peel each ply back to the cut. By this approach, a step-lap joint is produced, as depicted in Fig. 12.3.

Fastener hole formation is required for bolted repairs; in this case, the metal patch is used as a template to ensure correct hole alignment. Since the graphite/epoxy skin cannot usually be fully supported during this operation, very slow drill feed rates are required to avoid delamination and splintering damage during the breakthrough on the blind face.

Surface Treatment for Bonding

Graphite / epoxy patch. The most effective treatment for preparing preformed graphite/epoxy patches for bonding is grit blasting (usually with aluminum oxide grit) to remove contaminated surface matrix material. Another approach, which simplifies field application, relies on the use of a peel ply during formation of the patch. Peel ply is a layer of woven nylon cloth incorporated into the surface of the composite during manufacture. Prior to bonding the patch, the nylon is peeled off, exposing a clean surface ready for bonding. However, it is generally agreed that this procedure produces bonds inferior to those effected by grit blasting. This is because the grid-like nature of the peeled surfaces encourages air entrapment; there is also the danger of small amounts of the ply remaining on the surface.

By far the most effective procedure to avoid these difficulties and the danger of subsequent contamination is to cocure the patch and adhesive.

Graphite / epoxy parent laminate. The only options for surface treatment of the parent laminate are abrasion or grit blasting.

Titanium alloy. Metals generally require a much more elaborate surface treatment than the organic matrix composites. For durable bonds, it is usually necessary to form a stable oxide film (of the required morphology) by chemical etching and/or anodizing. In addition, the surface is usually coated with a corrosion-inhibiting primer. This has the extra function of protecting the metal part prior to bonding and is thus important for patches, which must be stored for field application. Chemical treatment under field conditions is best avoided where possible. Titanium alloys are usually surface treated with a proprietary etchant (such as Pasa Jell 107) based on a mixture of hydrofluoric and nitric acids in an aqueous medium, which is formed into a gel for field application.

Adhesives

Structural film adhesives are employed for applications where a high level of strength recovery is required. They generally consist of a precatalyzed modified epoxy such as epoxy-nitrile or, for high-temperature applications (60–100°C), an epoxy phenolic, supported by a polymer or glass fiber mat or woven cloth. More recently, similar materials have also become available in paste form. Provided that the curing conditions can be achieved and that moisture does not pose serious problems, film adhesives are preferred for repair applications. A serious disadvantage of the precatalyzed adhesives, particularly for repair applications, is that the cure reaction will slowly begin to occur—even under refrigeration to −20°C, life is only 6 months or so. Where this storage and replacement situation cannot be tolerated and lower mechanical properties can be accepted, two-part epoxy paste adhesives are employed since these have no storage problems and are also simpler to process.

Injection adhesives or resins do not have the same requirements for high peel strength as the adhesives used for bonding patches. However, they are required to have low viscosity and to suffer low shrinkage on cure. Generally, two-part casting or potting epoxies are used.

Curing Procedures for Field and Depot Repairs

Heat and pressure are required to cure the adhesive and obtain a uniform nonporous adhesive layer. Under field or depot conditions, these cure requirements are most simply satisfied with a vacuum bag arrangement, as illustrated in Figs. 12.5, 12.6, and 12.8. Although simple vacuum bag arrangements are capable of providing pressures of only 1 atm, this pressure is usually quite adequate if the patch mates well with the parent material; this condition is easily achieved by cocuring the patch and adhesive. However, higher pressures, if required, can be obtained by means of an oversize caulplate placed over the patch in the bag assembly.

Heat may be applied internally (as shown in Fig. 12.5) by encasing a heater element under the bag (usually an electrical resistance wire embedded in silicon rubber). Alternatively, a reusable combined vacuum bag and heater blanket may be employed, consisting of silicon rubber with built-in heater wires; this type of arrangement, although apparently quite attractive, has not proved to be very reliable. Heat to effect the cure may also be applied externally, for instance, in an oven (if the component is removable and of a suitable size) or by heat lamps, as shown in Fig. 12.8. Heat lamps are a very versatile method of applying heat, through nylon or Teflon film vacuum bags; however, it may be difficult to obtain an even heat distribution by this procedure and it is quite common to seriously overheat one spot while leaving nearby areas underheated.

The simple vacuum bag procedure suffers from several major drawbacks, all of which are associated with the low pressure that may be created in some regions inside the bag; these include: (1) the entrapped air and volatile materials in the resin matrix and adhesive may tend to expand under the reduced pressure, leaving large voids in the cured resin; (2) moisture absorbed in the graphite/epoxy parent laminate may be evolved more easily under reduced pressure and enter the adhesive, producing voids (and possibly interfering with the cure mechanism); and (3) air may be drawn into the bond region through any porosity in the parent material, producing voids in the patch system (the reduced pressure inside a honeycomb panel may cause the panel to collapse). Thus, although the vacuum bag procedure is reported to yield good results, its use has dangers, particularly for critical repairs. A safer alternative is to use pneumatic or mechanical pressure. The problem here is to arrange for the resulting loads to be reacted out. If they cannot be reacted out by the surrounding structure, vacuum pads or adhesively bonded anchor points may be employed.

12.5 DESIGN OF BONDED REPAIRS

General Considerations

Repair designs are most simply based on the "equivalent joint" (Fig. 12.3) obtained by considering a section taken through the damaged region. A more detailed and realistic analysis would need to consider the effective stiffness of the patched region and its influence on the surrounding structure; however, this aspect is beyond the scope of the present discussion.

Generally, the patch is chosen to match the strength and stiffness of the parent material. The strength of the joint (repair) may be designed to exceed by some margin (e.g., 20–50%) the allowable ultimate strain of the degraded parent material (after absorption of equilibrium moisture content) or, much less stringently, the allowable ultimate residual strain of the degraded parent material in the presence of strain concentrators such as holes or representative damage. If feasible, the first approach is most desirable, since it ensures that the adhesive bond cannot be stressed to failure, even under the most severe aircraft operating conditions and, further, it allows a reserve in strength for disbonds and defects. However, for most practical purposes, it should be quite satisfactory to design for the reduced strain levels.

For example,[5] the allowable ultimate residual strength for a $[(\pm 45°/0°/90°)_2]_s$ 16 ply (quasi-isotropic) laminate with 1% absorbed moisture and tested at 120°C, after representative fatigue loading, is about 9000 microstrain in tension and about 7500 microstrain in compression. Based on a nominal strain concentration factor of three for fastener holes, the reduced allowables would be about 3000 microstrain tension and 2500 microstrain compression. However, most design is currently based on an allowable of 3000–4000 microstrain, which is the practically determined allowable ultimate strain for laminates with fastener holes or impact damage, particularly under compression loading conditions.

External Patches

The external patch repair can be modeled as a half of a double-lap joint provided sufficient support is provided by the substructure to overcome bending effects. The following analysis is based on that given in Ref. 10 and initially requires a shear stress/shear strain curve for the adhesive at representative service conditions as shown in Fig. 12.9a for a typical film adhesive. It is assumed that (1) the patch is tapered at its ends to reduce peel stresses (very important for patches thicker than about eight plies or about 1 mm); (2) the patch has equal stiffness and thermal expansion coefficient to the parent material; and (3) the hole being covered by the patch is not tapered—this is a conservative assumption because any tapering improves the load-carrying capacity of the joint. Then, the maximum load-carrying capacity of the joint, based on an idealization of the stress strain behavior of the adhesive (Fig. 12.9b), is given by

$$P = 2\left[\eta \tau_p \left(\tfrac{1}{2}\gamma_e + \gamma_p\right) Et\right]^{\frac{1}{2}} \tag{12.1}$$

where τ_p is the effective yield stress of the adhesive, γ_e and γ_p are the elastic strain to yield and the plastic strain to failure, respectively, η is the adhesive thickness, t the thickness of the patch (and the parent material), and E its

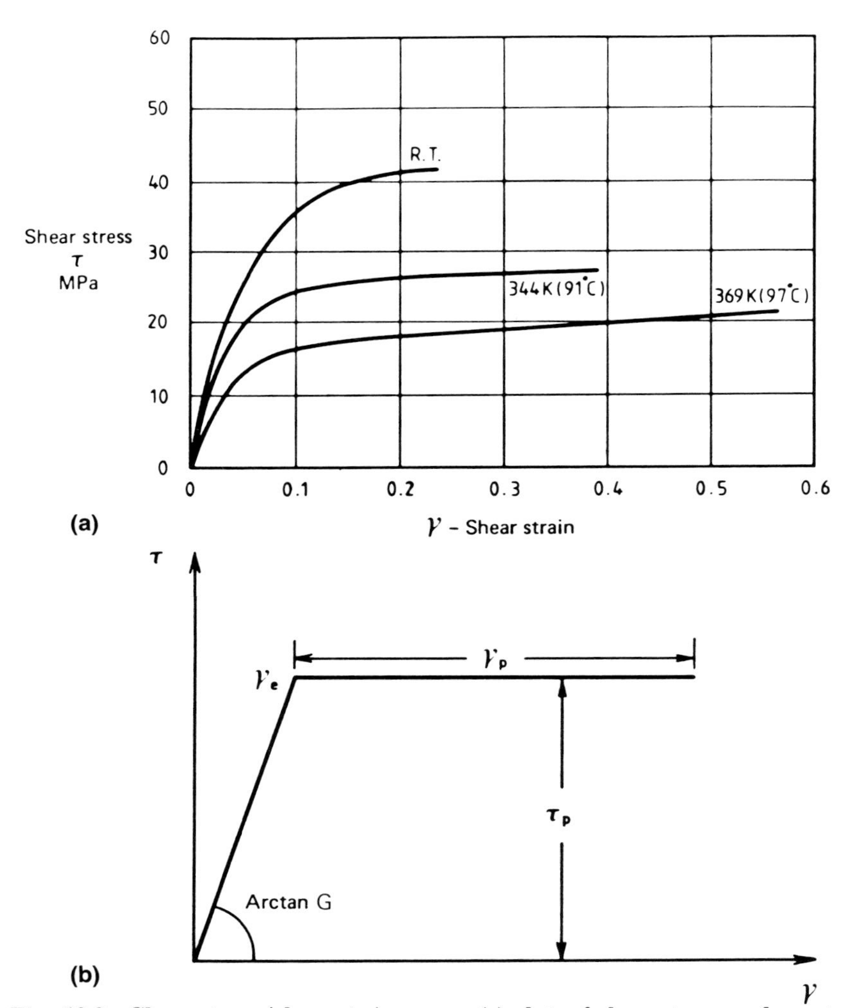

Fig. 12.9 Shear stress/shear strain curves: (a) plots of shear stress vs shear strain at various temperatures for adhesive FM 300 after exposure to moisture (results obtained from a thick adherend test); (b) idealization of a typical shear stress/shear strain curve for analytical purposes: the aim is to have an area under curve $\tau_p(\frac{1}{2}\gamma_e + \gamma_p)$ similar to that under the real curve.

modulus. As an example, taking (for hot/wet conditions),

$\tau_p = 20$ MPa
$E = 72$ GPa (typical for graphite/epoxy laminates employed in aircraft)
$t = 1.5$ mm (12 plies)
$\gamma_e = 0.05$
$\gamma_p = 0.5$
$\eta = 0.125$ mm

gives $P = 0.75$ kN/mm

The allowable load (per unit width) in the patch or parent material P_u is given by

$$P_u = Ee_u t$$

where e_u is the allowable ultimate failure strain of the composite. Taking this as 4000 microstrain gives

$$P_u = 0.43 \text{ kN/mm}$$

Thus, for the chosen ply thickness in this example, the load capacity of the bonded joint appears to be well above that allowable for the parent material. However, a safe margin is not obtained for external patch repairs of laminates above about 16 plies thick.

If it is considered that the strength of the repair is adequate, the next step is to determine the overlap length. The total patch length is twice this plus the diameter of the hole. The minimum design overlap length (ignoring the length of the taper) is given by

$$l = \left(\frac{P_u}{\tau_p} + \frac{2}{\lambda} \right) \times \text{safety factor} \tag{12.2}$$

where λ (the elastic strain exponent) is given by $\lambda = (2G/\eta Et)^{\frac{1}{2}}$ and G (the adhesive shear modulus) by τ_P/γ_e.

For the present example, using a safety factor of two to provide tolerance to damage, such as voids and minor disbonds, gives l as about 60 mm. Thus, assuming a hole diameter of 50 mm, the total patch length would be 170 mm.

Flush Repairs— Single-Scarf Configuration

Simple analysis. If the patch matches the parent material in stiffness and expansion coefficient, simple theory gives

$$\tau = P \sin 2\theta / 2t \quad \text{and} \quad \sigma = P \sin^2\theta / t \tag{12.3}$$

where τ and σ are the shear and normal stresses, respectively, acting on the adhesive and θ the scarf angle. At small θ, the normal stress σ is negligible. The required minimum value of scarf angle θ for an applied load P can be obtained from the following, taking τ_p as the peak shear stress:

$$P = Ee_u t = 2\tau_p t / \sin 2\theta \tag{12.4}$$

Thus, for small scarf angles, the condition for reaching the allowable strain e_u in the adherends is

$$\theta < \tau_p / Ee_u \text{rad}$$

Taking e_u as 4000 microstrain, τ_p as 20 MPa, and E as 72 GPa gives

$$\theta \leq 3°$$

Thus, taking the laminate thickness as 13 mm, the minimum length of the scarf is about 250 mm, which (for a hole size of 100 mm) gives a total patch length of 600 mm.

Repairs to laminates. The above simple theory is based on the assumption that the shear stress in the adhesive layer is constant. However, if the patch or parent material varies in stiffness with thickness—as does a composite laminate—the shear stress can no longer be taken as constant. In addition, when a composite patch is produced, its edges are stepped rather than scarfed. However, since the steps are very short, the shear stress in the adhesive may be taken as constant on each step. Thus, the repair is essentially a single (supported) step-lap joint, such as shown in Fig. 12.3, with very short steps.

The distribution of shear stress in the adhesive can be approximately obtained from the following simple analysis, based on a simplification of that given in Ref. 5, which ignores the shear lag effects and yielding in the adhesive.

From load equilibrium on each ply step of length Δx, it follows that

$$\tau = \Delta P / \Delta x$$

If it is assumed that the load increment ΔP on each step is proportional to the relative stiffness of the ply, such that for any layer

$$\frac{\Delta P}{P} \cong \frac{\text{stiffness of ply}}{\text{total stiffness}}$$

then τ varies through the laminate thickness approximately as does the ply stiffness. Thus, for a $[(0°/\pm 45°/90°)_2]_s$ laminate, the stiffnesses are in the ratio

$$1(0°) : 0.23(\pm 45°) : 0.07(90°)$$

This shows (on the basis of the above reasoning) that very high shear stresses occur on the 0° plies. These can lead to shear or, more likely, peel failure on the outer plies, unless they are relieved (for instance, by serrating the ends of the plies as described in Ref. 5). Serrations reduce the effective stiffness of the end of the ply and thus act in a similar way to tapering in an external patch.

An approximate estimate of the required step length Δx can be made by assuming the adhesive is stressed to its shear yield stress τ_p and each of the plies is loaded individually; then

$$\Delta x = \frac{E_p e_u}{\tau_p} \times \text{ply thickness} \qquad (12.5)$$

where E_p is the ply stiffness (typically around 120 GPa for the 0° plies). Taking e_u as 4000 microstrain gives a step length of about 3 mm. The length for the ±45 and 90° plies could be shorter, but in practice, for convenience, would probably be made the same length. Thus, for a laminate 100 plies (13 mm) thick, the scarf length is 300 mm.

12.6 CERTIFICATION OF REPAIRS

The question of certification must be addressed when repairs are made to critical structures. When sufficient experience has been gained with the design procedures, certification may be satisfactorily based on a stress analysis of the patched structure, together with data on materials allowables, obtained from tests on environmentally conditioned specimens of the various materials. However, at this stage, it is unlikely that all of the important factors have been fully appreciated (for instance, the influence of bending due to load eccentricity on the compression buckling strength and the long-term effects such as creep and stress relaxation on the adhesive). Consequently, at the present time, the design information must be supplemented by mechanical tests on the structural details representing the patched region.

A reasonable test procedure would be to expose such details to moisture and other relevant environmental agents for a prolonged period (at elevated temperature to aid diffusion) and then test under static loads to ensure that the structure will withstand ultimate design loads, test at temperature under cyclic loading (mission related) for, say, three lifetimes (or possibly less allowing for life consumed), and subsequently test under static loads to ensure that the ultimate residual strength has not fallen below the allowable level.

References

[1]"Adhesive Bonded Aerospace Structures," AFML-TR-77-206, AFFDL-TR-77-139, 1977.

[2]"Structural Sandwich Composites," MIL-HDBK-23A, U.S. Department of Defense, Washington, DC, 1968.

[3]Lubin, G., Dastin, S., Mahon, J., and Woodrum, T., "Repair Technologies for Boron Epoxy Structures," Paper presented at 27th Annual Technical Conference, Reinforced Plastics/Composites Institute and Society of the Plastics Industry, 1972.

[4]Studer, V. J. and La Salle, R. M., "Repair Procedures for Advanced Composite Structures," Vols. 1 and 2, AFFDL-TR-76-57, 1976.

[5]Myhre, S. H. and Beck, C. E., "Repair Concepts for Advanced Composite Structures," *Journal of Aircraft*, Vol. 16, Oct. 1979, pp. 720-728.

[6]Myhre, S. H. and Labor, J. D., "Repair of Advanced Composite Structures," *Journal of Aircraft*, Vol. 18, July 1981, pp. 546-552.

[7]Watson, J. C. et al., "Bolted Field Repair of Composite Structures," McDonnell Aircraft Co., Rept. NADC-77109-30, 1979.

[8]Myrhe, S. H., Labor, J. D., and Aker, S. C., "Moisture Problems in Advanced Structure Repair," *Composites*, Vol. 13, No. 3, 1982, pp. 289-297.

[9]Augl, J. N., "Moisture Transport in Composites during Repair Work," *Proceedings 28th National SAMPE Symposium*, 1983, pp. 273–286.

[10]Hart-Smith, L. J., "Analysis and Design of Advanced Composites Bonded Joints," NASA CR-2218, 1974.

13. AIRCRAFT APPLICATIONS

13.1 INTRODUCTION

As has already been described in Chap. 1, composite materials, especially graphite/epoxy, are being used to a significant extent in present-day aircraft and all signs are that this use will increase. Just to recapitulate, the major current applications of composite materials are listed in Table 13.1. In this chapter, the nature of some of these applications will be discussed further and an indication given of what seem to be the evolving general design rules. Also, some matters that need special attention when using composite structures (such as lightning and erosion protection) are noted. Brief mention is made of composite applications in helicopter construction.

For further information about particular applications of composite materials on specific aircraft, see Refs. 1–6.

13.2 TYPICAL COMPOSITE CONSTRUCTIONS

Skin Structure for Wings, Tails, and Control Surfaces

At the present time, the most common application of composites in aircraft structures is for the skin of wings, tails, and control surfaces. Generally, such skins are made in the form of "monolithic" laminates, i.e., without discrete stiffeners such as stringers on their inner surfaces. Orthotropic lay-ups are generally used and these can be readily analyzed by the stressing procedures described in an earlier chapter.

Consider first the skin of a main wing box for a straight-wing aircraft. This is primarily required to carry direct stresses in the spanwise direction due to the wing bending more or less as a cantilever beam and shear stresses caused by the wing twisting (Fig. 13.1); there will also be generally smaller direct stresses in the chordwise direction due to fore-and-aft bending. The simplest design approach is to use a laminate pattern (defined with respect to a reference axis in the spanwise direction) consisting of 0° plies to carry the spanwise direct stresses, ±45° plies to carry the shear stresses, and 90° plies to carry the chordwise direct stresses (Fig 13.2). Naturally, the number of plies of each orientation depends on the specific application. However, as an example, according to Ref. 2, the basic pattern for the F/A-18 wing skin comprises 46% of 0° plies, 50% of ±45° plies, and 4% of 90° plies. The required thickness of skin usually decreases markedly as one proceeds from the wing root to its tip. This is something that can be easily achieved using composites; it is necessary only to cut successively shorter plies prior to lay-up. Thus, a wing skin might be 100 plies thick at its root, but less than 20 plies thick at its tip.

The situation is much the same for the skin of a vertical tail (where the spanwise direction is, of course, upward) and a fixed horizontal tail. The laminate patterns are likely to be broadly similar to that described above for a wing. However, fighter aircraft often have all-moving horizontal tails ("stabilators" or "tailerons") and then, in order to provide adequate torsional strength and stiffness, it may be necessary to have a higher percentage of $\pm 45°$ plies, even up to, say, 80%. Also, the required skin thickness for a horizontal tail will generally be rather less than that for a wing and, at the tip, only a few plies may be required.

For a control surface hinged to a wing (Fig. 13.3), there is likely to be, relatively speaking, a much larger amount of chordwise bending and, consequently, its skin may be expected to contain a larger proportion of 90° plies.

Special reference should be made to the X-29 forward-swept wing demonstrator aircraft where the graphite/epoxy wing skins are made as nonorthotropic laminates. The laminates have been designed so that the interaction between the direct loads and the shear strains [cf. Eqs. (7.12) in Chap. 7] reduces the wing twist to an acceptable level.

Metal Substructures for Components with Composite Skins

Composite skins are widely used with some form of metal substructure. This is the case, for example, in the F/A-18 wing and in the F-14, F-15, F-16, and F/A-18 tail units. The substructure usually comprises aluminum alloy members (spars and ribs), but sometimes titanium alloy members are used. The F/A-18 main wing box has five aluminum alloy spars, but these have relatively small flanges so that the composite skin carries most of the

Table 13.1 Aircraft Applications of Composite Materials

Aircraft	Application
F-14	Boron/epoxy horizontal tail skins
F-15	Boron/epoxy horizontal and vertical tail skins
F-16	Graphite/epoxy horizontal and vertical tail skins and control surfaces
F/A-18	Graphite/epoxy wing skins, horizontal and vertical tail skins, speed brake, and control surfaces
AV-8B	Graphite/epoxy wing (skin plus substructure), horizontal tail skin, forward fuselage, and control surfaces
X-29	Graphite/epoxy wing skins
Boeing 757 and Boeing 767	Graphite/epoxy control surfaces, graphite-aramid/epoxy fairings, cowlings, etc.,
Lear Fan 2100	"Almost all" graphite/epoxy structure

bending loads. A generally similar type of construction is used in the F-16 vertical tail. The horizontal tails of fighter aircraft are commonly of small thickness with an interior structure comprising a full-depth aluminum honeycomb core. For an all-moving tail plane, a metallic spar/spindle is used and there may be light front and rear spars as well. An example of this type of construction, taken from Ref. 5, is shown in Fig. 13.4.

When metallic spars and ribs are used, the composite skin is generally attached to them by mechanical fasteners. As described in Chap. 8, when the skin is graphite/epoxy and the substructure an aluminum alloy, either titanium or stainless steel fasteners are employed to prevent galvanic corrosion. (For the same reason, thin sheets of some electrically insulating material such as glass/epoxy are inserted between adjacent areas of the composite and the metal.)

If a wing is built in the form of two half-wings, as is the case with the F/A-18, the wing-to-fuselage attachment needs special consideration. Composites are not well suited to carrying high bolt bearing loads. Because of this, a titanium fitting, which is bonded to the graphite/epoxy skins, is used at the root of the F/A-18 wing. This has been described in Chap. 8 and it will be recalled that a stepped-lap joint is employed; see also Ref. 2. The titanium fitting is then connected to the fuselage by lugs in the usual fashion. Broadly similar bonded joints are used for the root fittings of the F-14, F-15, and F/A-18 horizontal tails.

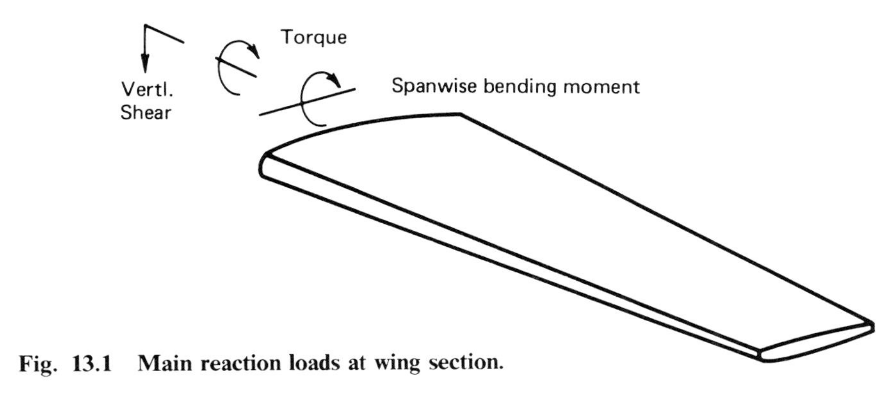

Fig. 13.1 Main reaction loads at wing section.

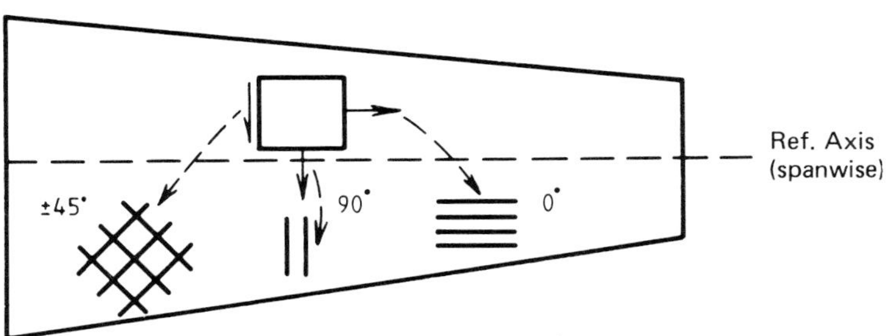

Fig. 13.2 Main ply orientations for wing skins.

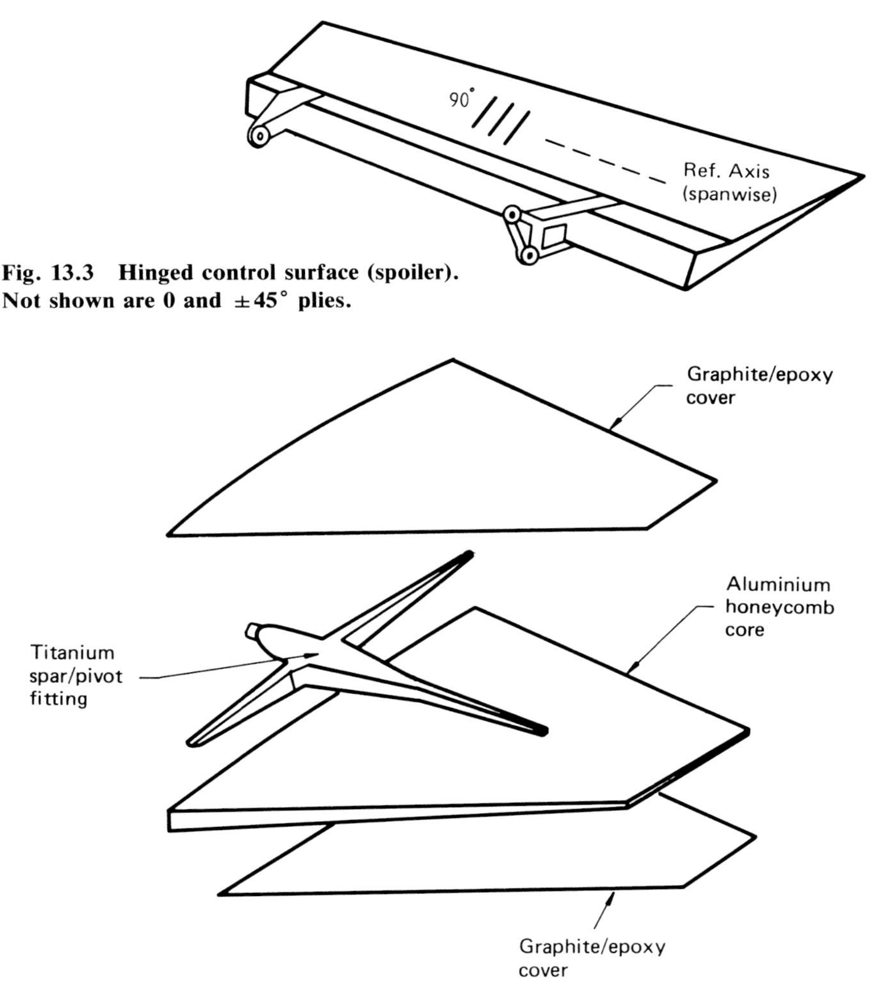

Fig. 13.3 Hinged control surface (spoiler). Not shown are 0 and ±45° plies.

Fig. 13.4 Representative composite all-moving horizontal tail plane construction (front and rear spars not shown).

All-Composite Wings

The AV-8B wing is a virtually all-composite structure; as well as having graphite/epoxy skin, it also has a graphite/epoxy substructure. The AV-8B wing, which incidentally is made continuous tip-to-tip (rather than in two halves as in the F/A-18), has eight graphite/epoxy spars. Because the main function of the web of a spar is to carry the vertical shear loads, the web comprises mainly ±45° plies, although there may be a small percentage of 90° plies also (Fig. 13.5). In the AV-8B application, the spar webs are corrugated in the spanwise direction to give increased buckling strength; the results are the so-called "sine-wave" spars.

Table 13.2 General Design Practice[9]

Serial	Practice	Reason
1	Filamentary controlled laminates—minimum of three-layer orientations	To prevent matrix and stiffness degradation
2	$0/90/\pm 45$ laminate with a minimum of one layer in each direction	0° layers for longitudinal load, 90° layers for transverse load, $\pm 45°$ layers for shear load
3	A $+45°$ and a $-45°$ ply are in contact with each other	To minimize interlaminar shear
4	45° layers are added in pairs (+ and −)	In-plane shear is carried by tension and compression in the 45° layers
5	When adding plies try to maintain symmetry	To minimize warping and interlaminar shear
6	Minimize stress concentrations	Composites are essentially elastic to failure
7	$\pm 45°$ plies, at least one pair on extremes of laminate. However, for specific design requirements (applied moments) 0° or 90° plies may be more advantageous in direction of moments	Increases buckling (strength) for thin laminates; better damage tolerance; more efficient bonded splice
8	Maintain a homogeneous stacking sequence banding several plies of the same orientation together	Increased strength

13.3 DESIGN PRINCIPLES FOR COMPOSITE STRUCTURE

Naturally, the design principles for composite aircraft structures are still evolving, but the following gives an indication of what seems to be current thinking.

According to Ref. 2, the design strains at the ultimate load condition for the graphite/epoxy components of the F/A-18 wing have been restricted to 5000 microstrain in zones of low fastener bearing stresses; in regions of higher bearing stresses the strain is further restricted. (For a 50% 0°/50% $\pm 45°$ laminate of graphite/epoxy with a Young's modulus of, say, 80 GPa, 5000 microstrain converts into an ultimate design stress of approximately 400 MPa.) Ultimate design strains of between 4000 and 4500 microstrain are recommended in Ref. 7; the lower figure, namely, 4000 microstrain, seems to be a fairly widely used one. For a reasonably extensive discussion of the design allowables for a typical graphite/epoxy system, see, for example, Ref. 8.

Dastin and Erbacher[9] have given what they consider to be the general design principles for composite structure; these are reproduced (verbatim) in Tables 13.2 and 13.3.

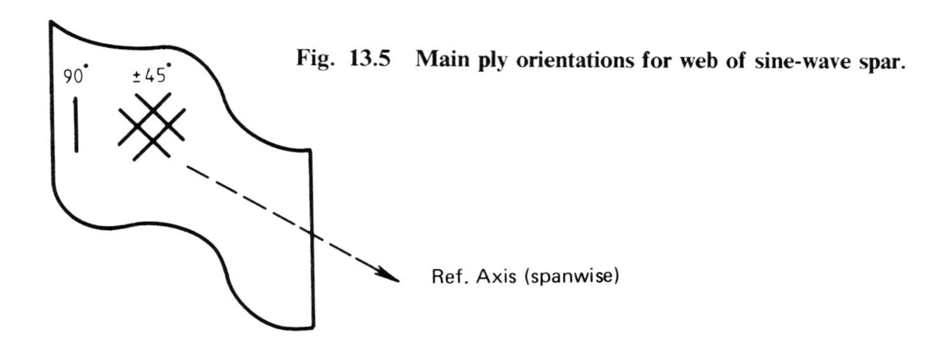

Fig. 13.5 Main ply orientations for web of sine-wave spar.

13.4 EROSION PROTECTION

The erosion resistance of composite materials, especially the resin matrix component thereof, is poorer than that of metals. Thus, for the leading edges of aerodynamic surfaces, it is usual to employ either an all-metal structure or else to bond a metallic layer over a composite skin.

13.5 LIGHTNING PROTECTION

Early composite aircraft structures used mainly glass or boron fibers. Since both these materials are essentially nonconductors of electricity, structures made of them are very susceptible to damage from lightning strikes; an account of the damage sustained by boron/epoxy structures under 100 and 200 kA simulated lightning strikes has been given by Clark.[10] The sort of protection schemes used for boron/epoxy components can be exemplified by that adopted for the F-14 horizontal tail.[1] There, 50 mm wide and 0.1 mm thick aluminum foil strips are bonded onto the composite skin at 100 mm centers. Fiberglass strips are located between the foil strips to preserve aerodynamic smoothness.

Carbon is, of course, a quite reasonable electric conductor. However, with graphite/epoxy structures, some form of lightning protection may still be incorporated; again, this may take the form of bonded metallic foil on the external surface of a component.

13.6 HELICOPTER APPLICATIONS

In the preceding parts of this chapter, attention has been limited to applications of composite materials in fixed-wing aircraft, but there is just as much interest in their applications to helicopters; see, for example, Refs. 11–14. While consideration has been given to the use of composites for many parts of a helicopter structure, there has been especial interest in two areas, namely, rotor blades and drive shafts.

All of the composite materials that have been discussed in this book, whether using glass, graphite, boron, or aramid fibers, have been seen as having distinct advantages for rotor blades (although boron is now probably ruled out on a cost basis). Pinckney[12] lists the following such advantages:

Table 13.3 General Design Practice for Joints[9]

Serial	Practice	Reason
1	3D edge distance and 6D pitch for bolted joints	Bearing strength
2	Design bonded step joints rather than scarf joints	More consistent results, design flexibility
3	Bonded joints, no 90° plies in contact with Ti	Reduction in lap shear strength
4	Bonded joints, ±45° plies on last step	To reduce peak loading
5	When adding plies use a 0.3 in. (7.6 mm) overlap in major load direction using a wedge-type pattern	Requires approximately 0.3 in. (7.6 mm) to develop strength

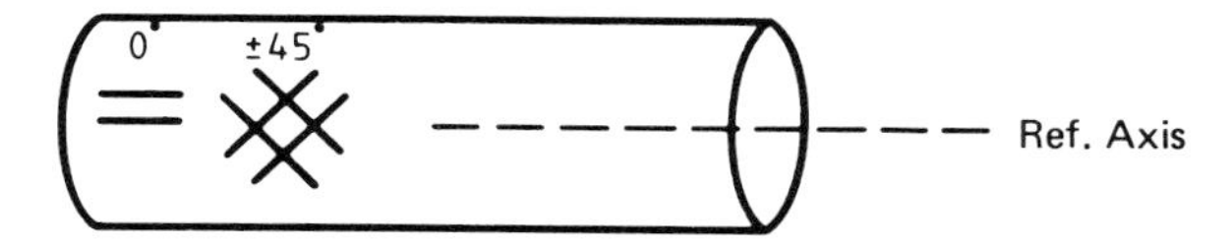

Fig. 13.6 Main ply orientations for a drive shaft.

(1) Improved aerodynamic efficiency is obtainable with composite blades because they can be readily manufactured in more complex airfoil shapes than are achievable with metal construction; Pinckney cites 20% increases in the lift-to-drag ratio from this cause.

(2) Although factors other than structural design may impose a minimum weight on a helicopter blade, there are still some circumstances where the weight saving that is generally achievable with composites can be utilized.

(3) According to Pinckney, the most significant advantage of the use of composites comes from the ability to tailor the dynamic frequencies and structural responses of the blade element to its operating parameters. The blade frequencies can be controlled by selecting appropriate laminate patterns to give the requisite bending and torsional stiffness.

Helicopters make an extensive use of connecting drive shafts, e.g., the usually long shaft to the tail rotor. Since a lack of stiffness is generally a potential problem with these, only the high-modulus composites (and not glass) are usually considered suitable for this application. The main function of a drive shaft is to transmit a torque; so, with a reference axis taken parallel to a generator of the shaft the laminate pattern is likely to comprise ±45° plies, with some 0° plies to provide increased bending stiffness (Fig. 13.6). (Actually, in the application described in Ref. 13, more 0° plies are used than ±45° plies; however, other applications described in Ref. 11 seem to use all ±45° plies.) According to Salkind,[11] a composite drive shaft may be easier to balance dynamically than an equivalent metal one, because of its higher internal damping.

References

[1]Forsch, H., "Advanced Composite Material Applications to F-14A Structure," *Composite Materials: Testing and Design (Fifth Conference)*, ASTM STP 674, 1979, pp. 30–39.

[2]Weinberger, R. A., Somoroff, A. R., and Riley, B. L., "U.S. Navy Certification of Composite Wings for the F-18 and Advanced Harrier Aircraft," AGARD-R-660, 1978, pp. 1–12.

[3]Huttrop, M. L., "Composite Wing Substructure Technology on the AV-8B Advanced Aircraft," *Fibrous Composites in Structural Design*, edited by E. M. Lenoe et al., Plenum Press, New York, 1980, pp. 25–40.

[4]Watson, J. C., "Preliminary Design Development AV-8B Forward Fuselage Composite Structure," *Fibrous Composites in Structural Design*, edited by E. M. Lenoe et al., Plenum Press, New York, 1980, pp. 41–62.

[5]Goodman, J. W., Tiffany, C. F., and Muha, T. J., "Structural Assurance of Advanced Composite Components for USAF Aircraft," AGARD-R-660, 1978, pp. 25–35.

[6]Vosteen, L. F., "Composite Aircraft Structures," *Fibrous Composites in Structural Design*, edited by E. M. Lenoe et al., Plenum Press, New York, 1980, pp. 7–24.

[7]Goodman, J. W., Lincoln, J. W., and Petrin, C. L., "On Certification of Composite Aircraft Structures for USAF Aircraft," AIAA Paper 81-1686, Aug. 1981.

[8]Ekvall, J. C. and Griffin, C. F., "Design Allowables for T300/5208 Graphite/Epoxy Composite Systems," *Proceedings of AIAA 22nd Structures, Structural Dynamics and Materials Conference*, Pt. 1, AIAA, New York, 1981, pp. 416–422.

[9]Dastin, S. J. and Erbacher, H. A., "Experiences with Composite Aircraft Structures," *SAMPE Engineering Series*, Vol. 25, 1980, pp. 706–715.

[10]Clark, H. T., "Lightning Protection for Composites," *Composite Materials: Testing and Design (Third Conference)*, ASTM STP 546, 1974, pp. 324–342.

[11]Salkind, M. J., "VTOL Aircraft," *Applications of Composite Materials*, ASTM STP 524, 1973, pp. 76–107.

[12]Pinckney, R. L., "Helicopter Rotor Blades," *Applications of Composite Materials*, ASTM STP 524, 1973, pp. 108–133

[13]Zinberg, H. and Symonds, M. F., "The Development of an Advanced Composite Tail Rotor Driveshaft," Paper 451 presented at 26th Forum of American Helicopter Society, 1970.

[14]"Special Composite Structures and Materials Issue," *Journal of the American Helicopter Society*, Vol. 26, No. 4, Oct. 1981.

14. AIRWORTHINESS CONSIDERATIONS

14.1 INTRODUCTION

As has been seen in the preceding chapters, the use of composite materials raises some problems that are different from those for metal aircraft structures. These problems, in turn, raise questions about specific airworthiness requirements for composite aircraft structures. At this stage, formal requirements are still being developed. As an example, consider the matter of the effect of the moisture/temperature environment on structural performance. Although the 1975 edition of the U.S. Military Standard on Aircraft Structural Integrity[1] states that the standard applies "to metallic and non-metallic structures," and although the U.S. Military Handbook[2] details general design procedures for composite structures, neither document specifies a procedure for allowing for environmental effects in the structural integrity program, including the static and fatigue tests on full-scale articles. According to Ref 3, written in 1980, airworthiness requirements for British military aircraft containing composite structure "exist in draft form" only. (Nevertheless, Ref. 3 gives quite a detailed account of key aspects of the airworthiness certification process for composite structures.)

Despite the fact that, to date, composites have been used much more widely in military than in civil aircraft, the U.S. civil authority has produced an advisory circular[4] setting forth an acceptable means of compliance with the "certification requirements of composite aircraft structures involving fiber reinforced materials, e.g., graphite, boron, and glass reinforced plastics."

The certification of the several U.S. military aircraft built to date that do contain significant amounts of composite structure seems to have been achieved basically by following the procedures for metal aircraft, but with additional requirements, sometimes of an ad hoc nature, agreed upon between the certifying authority and the manufacturer for the composite components; see Refs. 5 and 6.

It is not the aim of the present chapter to attempt to define a path to certification, but rather to give a general description of the type of approach that is evolving. As will be seen, two main problems are: (1) the proper allowance for environmental effects (moisture and temperature) in full-scale static and fatigue tests; and (2) the implications of the apparent increase in scatter in both the static and fatigue strengths of composite, as compared with metal, structures.

(With regard to scatter, as experience in the manufacture of composite structures has increased, there are indications that the scatter in static strength can be reduced to levels comparable with metals, i.e., with a coefficient of variation (defined as standard deviation/mean) of around 5%. On the other hand, it is generally agreed that the scatter in the fatigue strength of composites is much higher than that for metals, although, even here, it should be borne in mind that most of the composite fatigue data obtained so far have been for simple specimens; it is well known, and is consistent with extreme value statistics, that the fatigue scatter decreases as the size and complexity of the specimens increase.)

Brief mention is also made of some other special problems for composites.

14.2 DEMONSTRATION OF STATIC STRENGTH

General

The design procedure to establish the static strength of a metal aircraft generally involves, first, a detailed theoretical structural analysis (now almost always done using a finite element structural "model") and, second, a large amount of structural testing on specimens of varying degrees of complexity. These last can be conveniently, if somewhat arbitrarily, categorized as outlined in the following paragraphs. See Fig. 14.1.

(1) Coupon tests. These are tests on small, plain specimens used to establish the basic material properties (e.g., ultimate tensile strength). Enough tests are done on such specimens, so that "allowable values," which take account of statistical effects (i.e., scatter), can be established. Two allowable values are defined for any particular design quantity (e.g., ultimate tensile strength): the A-allowable, which is the value achieved by at least 99% of the population, at the 95% confidence level and the B-allowable, which is the value achieved by at least 90% of the population, at the 95% confidence level. An allowable value F_a for a quantity F is calculated from test data by a formula of the type

$$F_a = F_m(1 - K \times \text{CV}) \qquad (14.1)$$

where F_m is the mean test value, CV the coefficient of variation of the test values, and K a statistical parameter that depends on the relevant probability distribution (generally assumed to be normal), the number of samples, and whether an A- or B-allowable is required. For metals, it is usual to work with A-allowables and tables are given in Ref. 3 showing the reduction of an A-allowable below a mean value for various coefficients of variation and sample size.

(2) Structural detail tests. Structural details are also small specimens, but they contain typical, if elementary, structural features. Examples might be tension specimens with open or filled bolt holes, simple lap joint specimens, etc. If wished, it is possible to obtain allowable values for the strength of such details.

(3) Subcomponent tests. A subcomponent is a full-scale representation of a moderate-size region of a structure. Examples would be a compression

Serial	Type	Example ← L →	Typical dimension, L m
1	Coupon		0.1
2	Structural detail		0.2
3	Sub-component	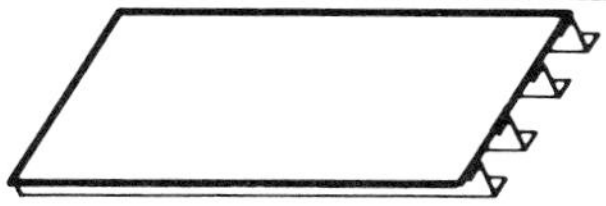	1.0
4	Component		3.0
5	Full scale aircraft	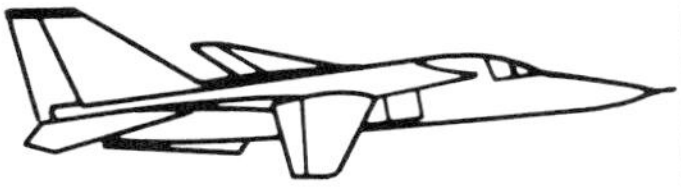	30

Fig. 14.1 Range of specimens in structural test program.

panel forming part of a wing skin, part of a major splice, etc. Generally, these specimens are already too large for it to be practicable to test enough of them to establish useful A-allowable values.

(4) Component tests. A component is a full-scale representation of a major region of a structure. An example might be a sizable length of the main wing box or a complete major joint. Here, only a very limited number of specimens (say one or two) may be available for testing.

(5) Full-scale article test. This is the test on a virtually complete aircraft structure. Almost certainly, only one test specimen will be available.

The above discussion has been from the design point of view. However, from the airworthiness point of view, it is the full-scale article tests that are seen as providing the key verification of static strength. Generally, there are two series of static tests on a full-scale article. The first is to establish that, at design limit load (DLL), no unacceptable deformations occur. The second is to establish that, at ultimate load, which by definition is 150% DLL, failure does not occur.

As has already been seen, some of the strength properties of composites are significantly degraded due to moisture/temperature effects and, possibly, although to a lesser extent, due to load cycle effects as well. It is clearly necessary that a composite aircraft have sufficient static strength even when it is in the most environmentally degraded form it will achieve in service. The question is, how is that to be demonstrated? The prospect of enclosing a complete aircraft in a humidity chamber, spending many months in moisture conditioning the structure, then possibly subjecting the structure to a certain amount of temperature and load cycling, before finally carrying out the static tests, is daunting. This is especially so if, because of scatter, it is unclear what inferences can be drawn from the results of a single test anyway.

Before going on, it might be mentioned that the matter of whether a composite structure should be given a certain amount of load cycling before being statically tested is rather contentious. The same argument can certainly be applied to a metal structure, but there it has usually been accepted that the static test can be carried out on a new structure.

Test Programs for Composite Structures

The same broad program outlined above for a metal structure is generally followed for a composite structure. Again, a finite element structural model is developed for the theoretical analysis and, again, tests will be carried out on specimens of varying complexity. The points of difference are discussed below.

(1) Coupon and structural detail tests. Since these two specimen types are of a generally small size, there are no major difficulties in conditioning them to whatever is considered an appropriate moisture level (typically of the order of a 1% weight increase). As indicated earlier, the specimens may, or may not be, subjected to some specified amount of spectrum load/temperature cycling (equivalent to one, or possibly two, lifetimes) before being statically tested to failure. It is usual to test according to the test matrix of Table 14.1.

Enough coupon tests are done on the main laminate patterns to establish the "allowable" values for each temperature/moisture combination. (Although the situation is by no means clear, there seems to be a trend toward accepting B-allowables for composite design—it being considered that, because of increased scatter, an unacceptably large amount of testing is necessary to establish A-allowables. The airworthiness implications of working with B-allowables, which are distinctly less conservative than A-allowables, are still to be assessed.) Enough structural detail tests are made to establish the worst environmental condition for each type of detail; then, enough tests are done for this worst condition and also for the RT/dry (base) condition, to establish allowable values for the detail in these two conditions.

Table 14.1 Matrix for Environmental Tests

	Temperature		
Moisture	Cold	RT[a]	Hot
"Dry"			
"Wet"			

[a] Room temperature.

As well as establishing design allowables, these tests also provide "knockdown" factors (i.e., reduction factors) for both the environmental effects and the variability. A comparison of, say, the "RT/dry" value and the "hot/wet" value of an allowable provides an environmental knockdown factor. Similarly, comparison of any mean value with its associated allowable value provides a variability knockdown factor.

As discussed in Chap. 7, because the stresses in a multidirectional laminate can vary markedly from ply to ply, allowable values are likely to be cited in terms of strain rather than stress.

(2) Subcomponent and component tests. The subcomponents and components selected for test will initially be based on the predictions of the finite element model (as, indeed, will have been the structural details). It will usually still be feasible, although time consuming, to moisture condition subcomponents and components prior to test. (Again, they may, or may not, be subjected to load/thermal cycles prior to test.) They are then statically tested to failure under the most severe environmental condition and the failure strain measured. These tests establish the mean values of ultimate strains in the environmentally degraded condition. It is also important to note the region and nature of the failure and to ensure that there are "structural detail" tests relevant to the region and that the same failure mode occurred there. If this is not the case, then such tests must be done. Then, assuming that the scatter in the subcomponent and component tests is the same as that in the detail tests (which is certainly open to criticism, but which is probably conservative), application of the variability knockdown factor to the mean values just determined gives allowable values for the full-scale structure in the environmentally degraded condition. In addition, if a component has not been tested with the environment included, then the environmental knockdown factor must also be applied.

(3) Full-scale article tests. The full-scale article is tested in the "RT/dry" condition. The main difference between this test on a composite and a metal aircraft is that, for the former, the structure is much more extensively strain gaged. This test serves to validate the finite element model; if the strain gage results show regions of high strain in areas where no component or subcomponents were tested, then it is necessary that such testing be done. Then, concentrating on the ultimate load test, the measured strains at 150% DLL are compared with the knocked down design allowables as established in test 2. If the measured strain exceeds the allowable value, failure is deemed to have occurred and some redesign is necessary.

Although there are uncertainties at various stages in the above test, the general approach seems reasonable. It can be seen that, for composite aircraft, virtually all development testing (on small and large specimens) becomes an integral part of the airworthiness certification.

However, it should be pointed out that the above is not the only approach to demonstrating static strength. One alternative sometimes proposed is to carry out the test of the full-scale article under ambient conditions (as above), but with the applied loads increased to allow for environmental effects. (The amount of the load increase is determined from specimen tests, much as already described.) Another alternative, of course, is to face up to a full-scale environmental test. Goodman et al.[7] have discussed these alternatives, which have various advantages and disadvantages.

Proof Tests

For some composite components in service, certain airworthiness authorities have required that every production component be given a proof test, generally to a load slightly in excess of the design limit load. In such cases, the components are given a thorough NDI both before and after test.

14.3 DEMONSTRATION OF FATIGUE STRENGTH

The situation with regard to demonstrating a satisfactory fatigue performance for composite aircraft structure is far from clear. The full-scale fatigue tests that have been carried out on current aircraft containing composites have generally been the same as would have been used for an all-metal aircraft, i.e., a test to N lifetimes (where N may be 2, 4, or whatever) in a normal environment. (Of course, most such aircraft are mainly metal anyway.) Again, the prospect of doing a fatigue test on a full-scale aircraft with the moisture/temperature environment fully represented is a daunting one. Also, sufficient data are not available on the fatigue of large composite structures to provide any real basis for selecting a scatter factor.

The full-scale test in a normal environment will certainly continue to be performed to verify metal structure and it may sometimes serve to reveal unsuspected problems with the composite structures. However, it seems that the main verification of the fatigue performance of the composite structure will be based on subcomponent and component testing in an appropriately humid environment, with the temperature cycling (especially the thermal spikes) accurately represented. Depending on what is eventually considered to be an adequate scatter factor for composite structural fatigue, these tests may or may not be performed in association with the static test program as previously described.

Because of the apparently large scatter in composite fatigue life that, if the same sort of fatigue test philosophy as has been used in the past for metal structures were applied, would seem to necessitate testing composite structures for excessively long periods (say, the order of 30 or more lifetimes), serious consideration is being given to adopting a different test procedure. For example, the composite taileron for the Tornado is being tested to only one lifetime, but an "enhanced" (i.e., more severe than actual) load spectrum is being applied.[8] (This sort of approach has previously been used for metal helicopter rotor blades.)

It should not be inferred from the above that the fatigue of composites is necessarily a cause for major concern; it is more a matter of there being difficulties in establishing a convenient test demonstration. A detailed discussion of the fatigue problem from the airworthiness point of view is given in Ref. 3.

14.4 DAMAGE TOLERANCE

An understanding of the damage tolerance requirements for composite aircraft structures is still in its early stages. Many uncertainties exist about the nature and extent of the damage that should be postulated in any damage tolerance requirements. Currently, delaminations seem to be the type of damage in which there is the most interest. Also, although this is very conjectural, there seems to be a line of thought that, in a large sheet structure, a laminate should be able to tolerate a delamination having a length dimension of the order of 20 mm without any serious degradation in its performance occurring over one lifetime.

14.5 A PROPOSED SCHEDULE OF ENVIRONMENTAL TESTS FOR COMPOSITE CERTIFICATION

Goodman et al.[7] have proposed an approach to the certification of composite structures, paying particular attention to the inclusion of environmental effects in the structural test program. (The matter of scatter is less thoroughly addressed.) The approach takes cognizance of the temperature seen by a particular component and, also, of its method of construction, namely, whether bolted, bonded, or cocured. The key features of the approach are summarized in Table 14.2, which has been taken verbatim from Ref. 7.

Table 14.2 Typical Full-Scale Structural Tests for 177°C (350°F) Curing Graphite/Epoxy Composite Safety-of-Flight Components[7]

<table>
<tr><th>Service Temperature</th><th>Assembly Method</th><th>Static Strength</th><th>Durability</th><th>Damage Tolerance</th></tr>
<tr><td>Less than 82°C (180°F)</td><td>Bolted, bonded, or cocured</td><td>Conventional room temperature static test</td><td rowspan="2">Conventional room temperature accelerated fatigue test</td><td>Room temperature accelerated damage growth tests</td></tr>
<tr><td rowspan="2">82–121°C (180–250°F)</td><td>Bolted</td><td>Room temperature increased loads</td><td rowspan="2">Accelerated worst-case moisture/temperature environmental damage growth tests</td></tr>
<tr><td>Bonded or cocured</td><td>Worst-case elevated temperature test, pre-conditioned for worst-case moisture absorption</td><td>Accelerated average moisture/temperature environmental fatigue test</td></tr>
<tr><td>More than 121°C (250°F)</td><td>Bolted, bonded, or cocured</td><td colspan="3">By individual agreement with procuring authority</td></tr>
</table>

14.6 FLAMMABILITY

In Ref. 4, requirements with respect to the flammability of composite structures for U.S. civil aircraft are given. Simply, these require that the existing levels of safety for metal aircraft should not be decreased by the use of composite structure. For aircraft exterior structure and engine compartment materials that are to be fire resistant, the following test is considered acceptable for demonstrating compliance. Two sheets specimens, each 620 × 620 mm (24 × 24 in.), are separately positioned perpendicular to a 1090°C (2000°F) flame produced by a modified oil burner consuming 9 liters (2 gal) of kerosene per hour. One of the specimens is of the composite material and has a thickness equal to the thickness of the component being evaluated; the other is of aluminum alloy with the thickness that would be appropriate if the component had indeed been made of aluminum alloy. The burner is positioned so that it takes the flame approximately 5 min to penetrate the aluminum specimen. The requirement is then that it should take at least as long for the flame to penetrate the composite specimen.

Although it is not strictly an airworthiness matter, it is convenient to refer here to a problem that, for a short time, caused substantial concern about the large-scale use of graphite fibers in aircraft. This was what sometimes was termed "the carbon fiber release problem" that arose in the following conceived situation. When a graphite structure is involved in a conflagration, very fine carbon fibers may be released into the atmosphere. These fibers may permeate electrical equipment in the vicinity of the conflagration and, since carbon is an electrical conductor, short circuits may result. (A scenario, sometimes described in the above connection, was of an aircraft fire in the neighborhood of an airport with the possible consequent long-term disruption of airport communication and navigational equipment.) An intensive study of this problem was undertaken in the United States[9] and the conclusion appears to have been that the problem is not nearly as serious as was first thought.

14.7 FORCE MANAGEMENT

The term "force management" refers to the procedures adopted when an aircraft enters service to ensure that it maintains a satisfactory standard of airworthiness during that service. A main activity here has always been the measurement of the loads experienced by an individual aircraft and then assessing the rate of consumption of its fatigue life. In practice, this has generally involved applying Miner's cumulative damage rule to the actual load spectrum and relating the results to the life established under the (usually different) fatigue test spectrum. There are two complications in this area for composite structures. First, as remarked in Chap. 9, there are some indications that Miner's rule may be quite unsatisfactory for composites; if that proves to be the case, then some alternative procedure will have to be developed so that the fatigue performance in service can be related to that demonstrated in the fatigue test. Second, because of the effect of moisture absorption on structural performance, it may well be important to monitor the moisture absorption in service "by tail number," along with the loads.

As also was indicated in Chap. 9, some monitoring of moisture absorption in service is already being done by measuring the weight increase in "traveler specimens," but a form of continuous monitoring would be much more convenient. Attention is being paid to identifying an appropriate physical property of a composite that can be both readily correlated with moisture absorption and can also be readily measured; for example, in Ref. 10, moisture-induced resistivity changes in graphite/epoxy were studied.

References

[1]"Military Standard, Aircraft Structural Integrity Program, Airplane Requirements," MIL-STD-1530A (11), U.S. Department of Defense, 1975.

[2]"Plastics for Aerospace Vehicles, Part 1, Reinforced Plastics," MIL-HDBK-17A, U.S. Department of Defense, 1971.

[3]Guyett, P. R. and Cardrick, A. W., "The Certification of Composite Airframe Structures," *The Aeronautical Journal*, Vol. 84, 1980, pp. 188–203.

[4]"Composite Aircraft Structure," U.S. Department of Transportation, Federal Aviation Administration, Advisory Circular, AC 20–107, 1978.

[5]Weinberger, R. A., Somoroff, A. R., and Riley, B. L., "U.S. Navy Certification of Composite Wings for the F-18 and Advanced Harrier Aircraft," *Certification Procedures for Composite Structures*, AGARD-R-660, 1978, pp. 1–12.

[6]Goodman, J. W., Tiffany, C. F., and Muha, T. J., "Structural Assurance of Advanced Composite Components for USAF Aircraft," *Certification Procedures for Composite Structures*, AGARD-R-660, 1978, pp. 25–35.

[7]Goodman, J. W., Lincoln, J. W., and Petrin, C. L., "On Certification of Composite Structures for USAF Aircraft," AIAA Paper 81-1686, Aug. 1981.

[8]Rogers, G. W., "The Fatigue Certification of the CFC Tornado Taileron," International Committee on Aeronautical Fatigue Conference, The Netherlands, 1981.

[9]"Assessment of Carbon Fiber Electrical Effects," NASA CP 2119, 1980.

[10]Belani, J. G. and Broutman, L. J., "Moisture Induced Resistivity Changes in Graphite Reinforced Plastics," *Composites*, Vol. 9, 1978, pp. 273–277.

Index

Acoustic emission, 186-187
Adhesives, 128-129, 209-210
Adhesive bonding, 115-131, 201-205
Aircraft:
  AV-8B (Harrier), 3, 181, 206, 220
  Boeing 727, 4
  Boeing 737, 4, 145
  Boeing 747, 2
  Boeing 757, 4, 218
  Boeing 767, 4, 218
  F-4, 2
  F-14, 2, 218, 222
  F-15, 2, 145, 218
  F-16, 3, 218
  F/A-18, 3, 127, 206, 217-219
  F-111, 2
  Lear Fan 2100, 4, 218
  Lockheed L-1011, 4
  McDonnell-Douglas DC-10, 4
  Mirage 2000, 4
  Rutan Voyager, 4
  Tornado, 3, 230
  X-29, 3, 218
Allowable values, 221, 226
Anisotropy, 64-65, 93
Aramid composites:
  applications, 4-5, 222
  fibers, 38, 43
  general, 1-2, 11
  hybrids, 71, 170
  insulation use, 139
  manufacture, 73, 88, 90, 91
  mechanical properties, 62-64
Autoclave, 79-82
Automated lay-up, 79
Average stress criterion, 159, 163

Ballistic damage, 154, 166
Barely visible impact damage (BVID), 155
Bearing strength, 135
Bending of laminates, 108-111
Bolted joints, 127, 131-139
Bolted patches, 205-206
Bonded joints, 115-131, 210
Boron composites:
  applications, 2, 6
  ballistic damage, 154
  boron/aluminum, 6, 11
  fibers, 38, 41-43
  general, 1-2, 7, 11, 145, 150, 177,
  manufacture, 73, 91
  mechanical properties, 27, 62-64, 165
  prepreg, 78
  repair, 198
Braiding, 90
Broadgoods, 44
B-stage, 55, 75
Buckling of fibers, 26-27

Carbon-carbon composites, 56
Carbon fiber composites, see Graphite composites
Carbon fiber release problem, 232
Certification tests, 215, 228-231
Chopped strand mat, 44
Cloth, 44, 93
Cocuring, 129
Coin tap test, 196
Compliance matrix, 65
Compression fatigue, 148
Compression strength, 134, 146
Contiguity coefficient, 19
Control surfaces, 194, 218, 220
Corrosion, 128, 138, 219
Cracks, 30-34, 161-165
Cumulative damage model, 24
Cumulative damage rule, 150
Curing, 82, 84, 210

Damage, 153-154, 195
Damage tolerance, 153-170, 231
Debulking, 79
Defects, 85, 119, 153, 194-195
Delaminations, 85, 106, 136-138, 154, 167-169, 196-197, 199, 200, 231
Design, 103, 115-119, 126-127, 131-136, 221, 223
Dielectric properties, 84-85
Double-lap joint, 117, 119-126
Double-scarf joint, 117
Double-stepped lap joint, 117
Drilling, 91, 138
Durability, see Fatigue

Eddy current NDI, 181
Elastic constants, see Moduli
Environmental effects, 50-51, 141-146, 149, 207, 228, 231
Epoxy, see Resins
Erosion protection, 222
Expansion molding, 84

Failure criteria, 111-113
Fasteners, 138-139, 219
Fatigue, 136-137, 147-150, 157-158, 230, 231
Fibers:
  aramid, 43
  boron, 41-43
  carbon, 39
  glass, 38-39

Kevlar, 43
volume fraction, 14, 60
Fick's law, 142
Filament, 43
Filament winding, 86-90
Flammability, 232
Flush patch repairs, 204
Force management, 232
Fracture surface energy, 30
Fracture toughness, 30-34, 164-165

Gel, 49, 80, 82
Glass composites:
applications, 2
fibers, 38-39
general, 1, 11
insulation use, 139
manufacture, 73, 75-76, 78, 88, 90-91
mechanical properties, 15-16, 62-64, 165
softening use, 131, 170
Glass transition temperature, 50, 56, 145
Graphite composites:
applications, 2-6, 217-222
damage, 155-156, 160-161, 166-169
environmental effects, 141, 144-146, 207, 231
fatigue, 147-150, 157-158, 230-231
fibers, 38-41
fracture toughness, 33-34, 165
general, 1, 11, 93
joining, 115, 123, 128, 131-132, 135-139
manufacture, 73, 82, 88, 90, 91
mechanical properties, 62-64, 97, 102
nondestructive inspection, 176-177, 180-183, 185-186
prepreg, 78
repair, 193-194, 198-199, 203-205, 207-210, 212

Helicopters:
drive shafts, 223
rotor blades, 88-90, 222-223
High modulus (HM) graphite, 40, 63
High strength (HS) graphite, 40, 63
Holes, 106-107, 133-134, 136-138, 156-157, 159-161
Honeycomb structures, 81-82, 194, 196, 200-203
Hybrid composites, 71, 170

Impact damage, 154-156, 170, 195
Ineffective length, 22
Injection repairs, 199
Interlaminar shear, 68, 86

Joints:
bonded, 115-131, 210
lap, 117, 119-126
mechanical, 131-139
scarf, 117, 126
stepped lap, 117, 126

Kevlar, see Aramid composites

Laminates:
codes, 107-108
manufacture, 74-86
orthotropic, 101-102, 104
quasi-isotropic, 103-104
symmetric, 98-99
Lay-up procedures, 79
Lightning protection, 222
Liquid impregnation, 74

Machining, 91
Macromechanics, 10
Mandrels, 87
Matched-die molding, 76, 82
Matrix materials, 1, 11, 47-57
Matrix volume fraction, 14
Mechanical joints, 131-139
Mechanical testing, 67-70
Metal matrix composites, 5-6, 11
Micromechanics, 9-10
Moduli:
laminates, 102-103
unidirectional composites, 12-16, 62-63, 94-95
Moisture absorption, 51, 134, 141-144, 200, 207-208, 225
Molds (tools), 79-83

Netting analysis, 87
Neutron radiography, 180
Nondestructive inspection (NDI), 86, 131, 175-191

Off-axis lay-ups, 28-30, 68, 95-97
Open-die molding, 75-76
Orthotropy, 66, 101-103

Paint, 146
Paint stripper, 146
Patch repairs, 199, 201-206
Peel ply, 129
Peel stresses, 125-126
Ply cutting, 79
Point stress criterion, 160, 163
Poisson's ratio, 15-16, 62, 102-103
Polyacrylonitrile (PAN), 39
Potted repairs, 201
PRD-49, see Aramid composites

Prepreg, 55, 59, 78-79
Pultrusion, 90-91

Quality control, 85-86, 130-131
Quasi-isotropy, 103-104

Release film, 78
Repairs:
  bolted, 205-206
  bonded, 201-204, 210-214
  criterion, 196-197
Resins:
  bismaleimide, 55
  epoxy, 1, 28, 48-52, 141
  PEEK, 56, 146
  phenolic, 53-55
  polyester, 52-53
  polyimide, 55
  thermoplastic, 56
  vinyl-ester, 53
Resin injection, 199
Rovings, 43
Rule of mixtures, 14, 16

Satellites, 4
Scarf joint, 117
Separator film, 80
Shear failures, 134-135
Shrink film, 77
Silicon rubber, 77, 84
Sine wave spars, 82-83, 220
Sizes, 45
Space shuttle, 4
Specific gravity, 63
Specific stiffness, 63
Specific strength, 63
Stepped-lap joint, 117, 127
Strain energy release rate, 32, 162-163
Strength, 19-30, 62-63, 111-113
Stress analysis, 104-105
Stress concentration factors, 106-107, 133, 159
Stress intensity factor, 32, 162-165
Stress resultants, 99-100, 109
Stress wave, 186
Surface treatment:
  bonding, 129-130, 209
  fibers, 45
Swelling, 145
Symmetric laminates, 98

Tailplane structures, 218
Tape, 45, 79
Tape laying, 79
Temperature effects, 50, 123, 145-146, 231
Thermography, 182
Thermoplastics, 56-57
Tows, 43

Ultrasonics, 86, 131, 171-180, 184, 196
Unidirectional ply, 9, 94-95

Vacuum bags, 79-81, 130, 203-205, 207, 210
Volume fraction, 14-15, 60

Warp, 44
Weft, 44
Weibull distribution, 24, 150
Whiskers, 6, 9, 38
Wing structures, 217-220
Woven roving, 44
Wrapping, 76

X-radiography, 86, 131, 176-177

Yarn, 43

Zero-bleed prepreg, 78